Alliant International University
Los Angeles Campus Library
1000 South Fremont Ave., Unit 5
Alhambra, CA 91803

Alliant International University
Los Angeles Campus Library
1000 South Fremont Ave., Unit 5
Alhambra, CA 91803

Alliant International University
Los Angeles Campus Library
1000 South Fremont Ave., Unit 5
Alhambra, CA 91803

The Anatomy of Bias

Alliant International University
Los Angeles Campus Library
1000 South Fremont Ave., Unit 5
Alhambra, CA 91803

The Anatomy of Bias

How Neural Circuits Weigh the Options

Jan Lauwereyns

The MIT Press
Cambridge, Massachusetts
London, England

© 2010 Massachusetts Institute of Technology

All rights reserved. No part of this book may be reproduced in any form by any electronic or mechanical means (including photocopying, recording, or information storage and retrieval) without permission in writing from the publisher.

MIT Press books may be purchased at special quantity discounts for business or sales promotional use. For information, please email special_sales@mitpress.mit.edu or write to Special Sales Department, The MIT Press, 55 Hayward Street, Cambridge, MA 02142.

This book was set in Syntax and Times by SNP Best-set Typesetter Ltd., Hong Kong. Printed and bound in the United States of America.

Library of Congress Cataloging-in-Publication Data

Lauwereyns, Jan, 1969–
The anatomy of bias : how neural circuits weigh the options/Jan Lauwereyns.
  p. cm.
Includes bibliographical references and index.
ISBN 978-0-262-12310-5 (hardcover : alk. paper) 1. Neural circuitry—Mathematical models. 2. Decision making—Physiological aspects. 3. Bayesian statistical decision theory. 4. Neural networks (Neurobiology)—Mathematical models. I. Title.
QP363.3.L38   2010
153.8'3–dc22
                                                                                                    2009021061

10 9 8 7 6 5 4 3 2 1

Contents

Prelude: A Double Inheritance, a Single Question   vii
The Beautiful but Ill-Defined Concept of Attention   viii
The Problematic Idea of a Book   ix
The Architecture of My Great Step   xiv
Before and After   xvii

1   Bayes and Bias   1
What Good Is a Mystery?   1
The Role of the Prior   7
Wrestling with Distributions   15
Time and the Measurement of Mind   28
Neural Signatures of Bias and Sensitivity   37

2   Wish Come True   49
What Is inside the Skin?   52
A Place in the Skull for All We Know   58
Ocular Movements in the Dark/Architecture and Morality   64
After Each Stroke the Vibrations   77
The Pursuit of Happiness   83

3   Fear Materialized   91
Round He Throws His Baleful Eyes   92
The Singularity of Restlessness   96
Snake Found Snuggled in Armchair   105
Who Is Afraid of Chocolate, Plush, and Manfred?   116

4   The Familiarity Fallacy   121
The Majority Rules   123
Implicit Associations and Flat Taxonomies   131
Something Straight from the Top   137
A Rose Is a Rose Is a Rose Is a Rose   142

5 The Proximity Trap 149
The Cheerleader Here and Now 150
Grouping by An-Pan-Man 155
The Trouble with Eternal Return 163
A Certain Surplus 170

6 Less Is More 173
A Bias toward Bias 174
Happy Accidents in the Evolution of Ideas 177
Monkey Stroop 181
A Toy Model of Competition in the Brain 190
Uniting with Language-Demons 195

7 Utopia—A World without Bias 199
Take Two 200
One Bias against Another 205
We Try to Overthrow Them 211
The Sentimental Duty of Punishment 214
My Own Desire to Eat Meat 218

Coda: The Pleasure of Prediction Error 223
What I Seek in Speech 224
Objects Represented to the Eye 226
This Makes the Blood-Vessels Flush 227
The Endeavor Has Just Begun 230

Bibliography 233
Index 255

Prelude: A Double Inheritance, a Single Question

From day one of my shadowy life in science there loomed large the inescapable figure of William James. Though my educational lineage as an experimental psychologist traces back most easily to Albert Michotte (of Leuven, Belgium) and then Wilhelm Wundt (of Leipzig, Germany), it was the hypnotic writing style of the American pioneer of psychology that first captured my attention, or started my imagination. It was not that I had read much of William James, or much of anything, really, but there were quotes floating around, little gems that kept surfacing time and again, in undergraduate courses and textbooks, in research articles, in papers presented at conferences, and even in texts far beyond safe academic waters. The quote that seemed most pervasive was this one, in which James (1950, pp. 403–404) claims everyone knows what attention is:

Every one knows what attention is. It is the taking possession by the mind, in clear and vivid form, of one out of what seem several simultaneously possible objects or trains of thought. Focalization, concentration, of consciousness are of its essence. It implies withdrawal from some things in order to deal effectively with others, and is a condition which has a real opposite in the confused, dazed, scatterbrained state which in French is called *distraction*, and *Zerstreutheit* in German.

I cannot remember where or when I first ran into this curious impromptu definition. I vaguely recall having a bit of a pedantic dispute with one of my teachers, Tom Carr (then at Michigan State University), about the grammar, particularly in the first sentence, with the "objects or trains of thought." Tom thought the mind took possession of either an actual "object" or else "a train of thought," but I insisted that such would be an odd pair of alternatives, a categorical error, almost, moving from a physical thing to a mental representation. Instead I suggested that the choice had to be between "an object of thought" and "a train of thought," with emphasis either on a static or a dynamic mental representation. (I'm not sure I explained my concern very clearly at the time; in any case, my subtle linguistic criticism would not have sounded very convincing in my thick Dutch accent.) Tom, of course, soon regretted inviting me to sit in on his graduate course on cognitive processes.

The Beautiful but Ill-Defined Concept of Attention

Today the little passage from *The Principles of Psychology* still retains some of its mysterious beauty for me, even though (or perhaps even more so because) I have now recited it dozens of times in my own undergraduate and graduate courses on brain and cognition at Victoria University of Wellington, New Zealand. (Dozens of times, yes, but no more than once or twice in each course, to be sure; it may be my mantra, but I don't hum it in class.) I like pointing out how funny it is that there should follow something after the rather confident observation that everyone knows what attention is. If we all know what it is, then why do we keep talking about it? I then usually go on to unpack the quote and briefly explore the ramifications of each of the mentioned characteristics of attention, with the opponent mechanisms of facilitation versus inhibition of information processing, the similarity to the spotlight metaphor of attention (often attributed to Posner, 1980, though Posner did not actually use the word "spotlight" in that paper), the notion that attention can be seen as the gateway to consciousness, and so on.

James sounds so hypnotic thanks to the leisurely cadence (and slightly idiosyncratic punctuation) with which he advances through the dense brush of concepts and abstract ideas, indulging, for instance, in a triplet of adjectives to specify a state ("confused, dazed, scatterbrained") which he then proceeds to push into the unknown again by associating it with exotic words, borrowed from French ("*distraction*," which, italicized, apparently does not simply mean "distraction") and German ("*Zerstreutheit*," which, roughly translated to an English neologism, would be "scatteredness" or, indeed, the scatterbrained state). The text is so lucid and lovable that it forgets to teach and becomes confusing, providing the ultimate image of what it was trying to say all along, namely, that we can know, and yet not really know, what attention is. This is precisely the point of departure for a science that wishes to address the core experience of being (or life) as we know it. William James called it "psychology," a term that no longer carries quite the same meaning as it did when he had its principles in mind. Psychology has become fractionated, and the various objects or trains of thought that James brought together in his *opus magnum* would in the present day find their offspring in the divergent fields of cognitive neuroscience, psychoanalysis of the post-Lacan variety, literary theory, philosophy of mind, evolutionary psychology, experimental economics, and several other lines of inquiry. Being a fan of James and the breadth of his vision, I will recklessly endeavor to scavenge materials from these various fields with the single aim of producing a coherent but open-minded account of attention, or bias versus sensitivity, or how the activities of neurons allow us to decide one way or another that, with a faint echo of Hamlet in the background, something appears to be or not to be.

In the meantime, it may be useful to situate the inspiring quote in its original context. It is buried deep in the first volume of the *Principles*, in chapter XI on the topic of, indeed, attention. James started the chapter with the observation that "psychologists of the English empiricist school" (by which he refers to Hume and Locke among others) had failed to give the concept of selective attention its fair due in their investigations; they had too easily made experience "tantamount to the mere presence to the senses of an outward order" (1950, p. 402), as if attention were merely a slave to its objects. This could not be the entire story, James reasoned (still on p. 402):

Millions of items of the outward order are present to my senses which never properly enter into my experience. Why? Because they have no interest for me. My experience is what I agree to attend to. Only those items which I notice shape my mind—without selective interest, experience is an utter chaos.

Here, we are dangerously close to circular reasoning, paradoxical thinking, or a form of thought that would have no place in science, other than as a boundary condition, or the ultimate void of meaninglessness. My attention is supposed to go to items that hold some interest for me, but how can I select such items unless I already know what they are? What use would I have for a selection mechanism that operates only on things that are known to me already? James seems to suggest that it is primarily a matter of mental hygiene; selective interest is needed to shape my mind and to prevent utter chaos. It would be a cognitive economy of sorts, and thanks to the organizational power of my attention, and the space it frees up in the toy model of the world inside my head, I would perhaps be able to think deeper, weigh different options and courses of action, and make better decisions. If this is true, the study of attention should receive an urgent and central place in the human effort to understand all things human; if it is indeed the core factor that stands between experience and chaos, then all areas of human life, from the most mundane to the most esoteric, would stand to benefit from a detailed examination of its characteristics, its conditions and implications.

The Problematic Idea of a Book

For better or worse, beautiful and ill-defined as it was, the topic of attention seemed an excellent place for an eager mind to enter the world of science.

I was eager enough, and so my mind did just that (enter the world of science) exactly there (on the topic of attention), though it took me a while, and a string of great teachers, before I realized it myself. In part, the delay in my vocation was due to something I like to think of as a benign multiple personality disorder, which is probably not so very uncommon, given the number of highly talented scientists I have met who are engaged in some other intellectual passion, artistic or musical. Two of my favorite examples are the expressionist painter–engineer–computational neuroscientist Minoru Tsukada and the baritone–psychophysicist–physiologist R. H. S. (Roger) Carpenter. One not-so-very-hidden hidden agenda of this book will be to present an example of radical cross-disciplinary interests that actually might be good for each other. Given that I have two left hands and a tendency to spill liquids, and knowing that I couldn't sing two notes to save my life, my own creative impulses had to be directed to the art of words, to poetry, prose, any form of writing that cared about how it said what it said, or even professed that saying something meant saying something in a certain way, and if you wanted to say it right, you had better consider how you were going to say what you wanted to say in the first place. Thus, I was condemned to literature. And in the literary arena I stumbled across William James again:

As we take…a general view of the wonderful stream of our consciousness, what strikes us first is [the] different pace of its parts. Like a bird's life, it seems to be made of an alternation of flights and perchings. The rhythm of language expresses this, where every thought is expressed in a sentence, and every sentence closed by a period. The resting-places are usually occupied by sensorial imaginations of some sort, whose peculiarity is that they can be held before the mind for an indefinite time, and contemplated without changing; the places of flight are filled with thoughts of relations, static or dynamic, that for the most part obtain between the matters contemplated in the periods of comparative rest.

I encountered this wonderful, lyrical passage from *The Principles of Psychology* (James, 1950, p. 243) in a collection of essays, entitled *The Language of Inquiry*, by the American poet Lyn Hejinian (2000, p. 121). Hejinian's own work is notorious—celebrated by some, including me, cast aside by others—for its relentless commitment to experimentation and literary invention; she was on the forefront of the so-called L=A=N=G=U=A=G=E movement in poetry, driven by a very diverse group of poets who emerged as leading new voices in California's Bay Area in the 1970s. Their one common feature was a shared meta-analytic interest in the medium of their art. Of these poets, Hejinian is probably the poet with the most "scientific" attitude. In fact, the William James quote that Hejinian used in her book is very telling about her prerogatives, demonstrating exactly what James said about attention and how we only take note of what holds an interest for us. For Hejinian, the sentence is the principle unit of language, or, stated in a formula,

POETRY = SENTENCE ART.

Hejinian's William James quote hit me like a boomerang. A few years ago I was sure I had had my fill of the intentionality of visual selective attention (the title of my PhD thesis). But if I had thought to escape the legacy of William James, the hard-core cognitive psychologists, the reductionist spotlight theorists of attention, then here was James with a vengeance, right in my beloved territory of experimental poetry. Hejinian's quote was on the mark in several ways, not just with respect to her own poetics and interests, or as an illustration of the near-paradoxical mechanism of attention, which gives us only what we had already, but also as a final vindication of my interpretation of the other quote, with the static versus dynamic modes of thought (the grammatical issue with the objects and trains). The new passage proved me right and Tom Carr wrong, I concluded, and so maybe I had a future as a William James kind of scholar after all.

Slowly, then, the idea of writing about my field of study started taking shape; perhaps to really approach the topic of attention and decision making in the way that best exploited my skills, feeding from my strengths and acknowledging my weaknesses, I would have to not only process and compile technical reports of scientific experiments in line with the culture of the profession but also seek to understand and be able to talk about the findings and observations in a language that would invite readers from outside the core circle of cognitive scientists to take part in the inquiry. Again, William James (1950, pp. 252–253), as (not entirely accurately) quoted by Lyn Hejinian (2000, p. 122), would be my guide:

The truth is that large tracts of human speech are nothing but signs of direction in thought.... These bare images of logical movement...are psychic transmissions, always on the wing, so to speak, and not to be glimpsed except in flight.

The original version by James had "psychic transitions," though I think I prefer Hejinian's "transmissions." But especially the last phrase invariably reverberates in my mind's ear whenever I reread the quote. The echo must needs be heard, "always on the wing, so to speak, and not to be glimpsed except in flight." I take it as an image of the extraordinary power of human speech, even when all that talking rather feebly seems to give "nothing but signs of direction in thought." These directions of thought are exactly the stuff that scientific progress is made on, and it is in the exercise of speech that we become aware of the psychic transactions, once again, "always on the wing, so to speak, and not to be glimpsed except in flight."

Writing about my field of study, then, might achieve something like this, giving a glimpse of the logical movements in flight, showing how the cognitive neuroscience of attention and decision making got to where it is at present, and looking ahead to where it might be going next. If James is right, and if my speech is to the point, this exercise—or the bird's eye view, to continue the metaphor—would be useful even to the experts who are defining the research programs. Which trends are important? Which kinds of experiments are viable? Which approaches are likely to yield new information? I think we need to think about such things every once in a while, before rushing off to more data collection:

> The role of the thinker is not so much to utter truths or establish facts, but to distinguish among the large population of true facts those that are important and relevant from those that are not.

The words are from Manual DeLanda (2004, p. 7) in his characterization of Gilles Deleuze's epistemology. They are appropriate in any context, I believe. DeLanda, by the way, is another fine example of a very productive cross-disciplinary spirit. He started out as an experimental filmmaker but found his way to philosophy through the work of the idiosyncratic French philosopher, who happened to share DeLanda's passion for cinema. DeLanda proceeded to make it his mission to save the useful bits of Deleuze's work for the science-oriented analytic philosophers who generally do not have much time for what is rather pejoratively termed "continental philosophy." It turned out Deleuze had said useful things about dynamics, chaotic systems, and so forth for even the most mathematically inclined biologist with a phobia for full sentences.

DeLanda's, and Deleuze's, point is that we sometimes need to take one step back and survey the journey traveled so far to find the best direction to move forward. One experiment easily leads to another, but does it really tell us something new, does it actually move to a more precise level of description, or does it merely provide another paraphrase of some already-established correspondence between one thing and another?

At first glance, the current publication culture in science does not seem to be very conducive to the kind of speech act that I wish to engage in with the present book. Researchers are under extreme pressure to publish boatloads of papers as quickly as possible, preferably but not necessarily in top journals. To accommodate this problem, over the last ten or twenty years new journals have sprung up like mushrooms in a damp forest. The number of new online journals in the last two or three years alone is simply staggering. The principle benefit is obvious; researchers can now easily archive massive amounts of data. The costs, however, are indirect, often ignored, and potentially disastrous for the entire scientific enterprise. The archived data are often of substandard quality, or even if they are good-quality data, the analyses have been rushed, the discussion is too summary, and the paper in general lacks sharpness and precision. Worse, the sheer volume of published reports makes it impossible for researchers to keep track of all relevant publications; there are vastly more papers than before, and virtually all of these papers pass to oblivion in a few years or even a few months. The net effect of publishing extensively is that you are expected to publish even more—funding agencies, administrators, and people on the sidelines expect more of the same. All of these factors combined create the clear and present danger that labor-intensive, long-term research programs become an economic liability for the individual researcher. Who would ever have the luxury of taking a half year off to write a book-length critical review? One monograph equals four or five technical reports that will forever remain unpublished.

Perhaps we simply need to stop worrying and do what we wish to do, or, as a fourteenth-century Buddhist priest in Kyoto suggested (Kenkō, 2001, p. 53), "A man who has determined to take the Great Step should leave unresolved all plans for disposing of urgent or worrisome business." Luckily, science is hosted by institutes and organizations that do provide researchers with some protection against immediate pressures that run counter to "the Great Step" of sustained, rational thinking. Throughout my work on this book, I have had the good fortune of being kept out of the rain by four such umbrellas. I gratefully acknowledge them here: Victoria University of Wellington, New Zealand, my academic home base, which gave me the opportunity to take an eleven-month research and study leave from June 2008 to April 2009; the Brain Science Research Center of Tamagawa University, Tokyo, Japan, the host institute that took me in as a Visiting Scholar during that time; the Japanese Society for the Promotion of Science, which awarded me a very generous Long-Term Invitation Fellowship for Research in Japan (award number L-08507); and The MIT Press, a strong early supporter of the project.

I would strongly like to encourage other researchers to take advantage of similar opportunities to step out of the usual routines of daily life in the lab and reflect on the big picture of our fields of study. It is only through such efforts that we can hope to escape the pitfall of hastily publishing too many papers containing too little information. Let us take time and think things through. In the words of Lyn Hejinian's wonderfully uncompromising poem "Happily" (2000, p. 385):

> Constantly I write this happily
> Hazards that hope may break open my lips
> What I feel is taking place, a large context, long yielding, and to doubt it would be a crime against it
> I sense that in stating "this is happening"
> Waiting for us?

We are waiting for us, indeed, to make this take place, "a large context, long yielding," providing "hazards that hope may break open my lips"—odd words that force some kind of shift: of perspective, if not of paradigm.

Of course, now the burden is on me to provide something substantial and new. How will I do that? Like Hejinian, I will use odd words in the hope of forcing some kind of shift, not of paradigm but of perspective. Though my topic stays very close to what William James thought we all knew, I will talk about "bias" and "weighing the options," or concepts that are not exactly synonymous with the more common terms "attention" and "decision making." Indeed, I will spend a good amount of time and effort pleading for a language that is as ergonomic and simple as possible to move from brain to behavior, and back again, in ways that allow researchers to be more precise than before about the correspondence between the things we do and the things that happen inside our head.

I may be a William James kind of scholar, but the problem with "attention" is that it is too vague—so there will be a desire for a touch of good old behaviorist thinking in the fabric of this book as well. The aim is to combine breadth of vision with conceptual minimalism. For this, "bias" will be my key concept.

The Architecture of My Great Step

What is bias exactly? Does it simply mean preference for one thing rather than another? What is the role of bias when we make judgments about objects and events, from simple, sensory categorization to complex, social assessment? How does bias emerge in neural circuits? My great step, the book I must now write, should give an integrative account of the structure and function of bias as a core brain mechanism that attaches different weights to various information sources, prioritizing some cognitive representations at the expense of others. The principal objective is to provide a critical analysis of this core brain mechanism in a single volume, offering an engaging and comprehensive narrative. The project is not to produce a textbook in the classical sense—it does not purport to replace, for instance, the authoritative 800-page reference work on *Memory, Attention, and Decision-Making* by Edmund T. Rolls (2008). Instead, my hubris will be to think I should seek to open up the research field to new or other modes of thought, believing that there may be ideas and knowledge to be gained from explicit efforts to rethink the data, applying a different perspective—here, with emphasis on the pervasive concept of bias.

The book is primarily aimed at the kinds of minds that would be ready to begin a postgraduate program in psychology or neuroscience, or perhaps the third or fourth year in the more demanding undergraduate programs, but I hope it will also attract more advanced readers from the same and neighboring research fields. As an authored book, speaking in a single voice and providing an integrative account, it promises to be an attractive point of entry to an otherwise amorphous and vast literature—a literature which may be quite daunting for anyone who is willing to invest time in reading one "content-heavy" book but simply does not have the time or the means to survey the various expansive bodies of literature on the topic of bias in decision making.

The book endeavors to be innovative in two major ways. First, I identify "bias" (rather than more noncommittal terms such as "selection" or "attention") as a core concept to study the neural underpinnings of action control and various forms of decision making. This allows researchers to apply existing computational tools and be relatively precise about how neurons contribute to the control of behavior. Second, I aim to provide explicit linking propositions with other contemporary analyses of decision making in an effort to stimulate mutual interest (and multidisciplinary discussion) among students and professionals from within and outside the area of cognitive neuroscience.

Chapter 1, "Bayes and Bias," develops the concept of bias on the basis of notions from Bayesian probability, signal detection theory, and current neural models of decision making. "Bias" here refers to the unequal weight given to different sources of information—as a result of preferences, prejudices, or other forms of selective processing. The aim is to translate the computational properties of bias into easily recognizable neural signatures. These signatures are introduced by means of concrete examples from the experimental literature. The chapter also includes a brief overview of the types of experimental paradigms that provide relevant data—this stands as a short "discourse on method" (inviting a brief detour on epistemology).

Chapter 2, "Wish Come True," focuses on how the reward factor is implemented in neural mechanisms for information processing. A distinction is made between anticipatory mechanisms that bias the neural system on the basis of reward probability and synergistic mechanisms that influence the quality and resolution of information processing (from perception to memory). This chapter also builds a bridge to the new field of "neuroeconomics" and perspectives from evolutionary psychology, as well as literary, philosophical, and psychoanalytic approaches to the concept of desire.

Chapter 3, "Fear Materialized," explores the flip side of the coin—negative motivation as a determinant of the priority given to different sensory inputs.

Important questions include the extent of overlap with the reward-oriented mechanisms, the possibility of opponent dynamics, and the similarity of the underlying (anticipatory versus synergistic) algorithms. I carefully examine previous reports of "automatic attention" evoked by potent emotional stimuli and suggest a reinterpretation with the concept of bias. The mechanisms of fear conditioning turn out to stimulate neural plasticity and produce structural forms of bias. Again, the discussions afford comparison with notions from neuroeconomics and evolutionary psychology, as well as more psychodynamic explorations of fear and aversion.

In chapter 4, "The Familiarity Fallacy," I examine the neural underpinnings of what is perhaps the most stereotypical image of bias in society—namely, racism, sexism, or any form of prejudice in which familiarity goes hand in hand with priority and beneficial treatment. Do such effects emerge inevitably from the structure of neural circuits, or can we divorce familiarity bias from prejudice and social inequality? If information propagation depends on learning and preestablished neural connections, then what are the behavioral implications? The ability to learn is an obvious prerequisite for flexible and well-adapted behavior, and this, in turn, raises the question of when the familiarity fallacy is really a *fallacy* (or where do benefits end and costs begin?).

Chapter 5, "The Proximity Trap," departs from the classic observation, first offered in the work of the German Gestalt psychologists, that spatial and temporal properties of information, such as proximity, similarity, and continuation, can generate perceptual grouping. In this chapter I revisit these phenomena from the perspective of bias in neural processing, along with other "history" effects that have typically not been associated with the literature on perceptual grouping (e.g., repetition priming). A dominant theme here is the autonomous nature of the perceptual organization, giving rise to "pop-out" phenomena and exogenous attention capture. A comparison is made with the physics of attractors and self-organization in chaotic environments.

In chapter 6, "Less Is More," I discuss the implications of the simple fact that individuals or organisms operate in a constrained "answer space"—they can only do so much. Bottleneck and other metaphors of selection have abounded in psychology and neuroscience alike, but I propose that the *reinforcing* aspect of selection has been underinvestigated. Horse race and other competition models are congruent with the notion that competition raises the stakes and thus could lead to an intensification of information processing, while at the same time the winner of the competition "takes all" (or will become the object of more extensive processing, such as being maintained in working memory or transferred to long-term memory). This might lead to a systems bias for bias, following a slogan such as "Less is more," so that selection and bias are applied even in situations in which it would in principle be possible to be noncommittal, weighing all information equally. Relevant discussions in this regard are those on the human mind's penchant for simplicity and its potentially damaging implications.

Finally, in chapter 7, "Utopia—A World without Bias," I conduct a thought experiment, trying to imagine what a world without bias would be like. From the wealth of evidence presented in the previous chapters it is quite clear that bias in neural processing is so pervasive as not to be eradicated by any means, nor would we necessarily benefit from the absence of bias (chapters 2 and 3 clearly connect bias with inclusive fitness and evolutionary success). Instead, we must focus on how to *control* bias—knowing when it can be used advantageously but also being alert to conditions that require us to work against bias. The opposition to bias, and the self-control that it implies, may be a crucial component in adaptive behavior of a "higher order," in terms of delay of gratification, long-term investment, and sacrifice for group or social benefit.

Before and After

If all goes well and my great step delivers on its promises, the book that has now been written will die a slow death in the best scientific tradition, being superseded by the work of a student or, better yet, the works (plural!) of students (plural!). The scientific investigation of bias in decision making is still in its infancy and will continue to benefit greatly from the immense rate of technological development. Only a half century ago, the British psychologist Donald Broadbent (1961, p. 302) wrote the following:

It seems likely that no conceivable observation can overcome the uncertainty about the position and velocity of an electron, since determining one disturbs the other. Events within the skull are not inherently unobservable: they are merely inaccessible to our present techniques. The proper analogy for the most speculative statements in psychology is not with, say, a proposition about velocity through the ether, where there is some doubt whether the proposition is meaningful. It is rather with propositions about mountains on the other side of the moon, which are perfectly meaningful but rather difficult to test. The mountains are as difficult to observe as the velocity, but the reasons for the difficulty are purely technological.

The statement is very much true today. In defiance of skeptics who have time and again declared the death of science in general, or the inherent inaccessibility of the relation between brain and mind in particular, the enterprise of cognitive neuroscience has all but made the most arcane phenomena of human consciousness tractable in the laboratory. The present book is a part of that enterprise and invites others to join in.

Of course, I did not just randomly pick a quote from Broadbent to broadcast the invitation. Donald Broadbent was one of a handful of scientists who helped shape the literature on selective attention during the 1950s and 1960s in cognitive psychology; his well-known filter model was, in part, inspired by the work of William James. Broadbent was the most frequently cited scientist in my own master's thesis, and I have a vivid memory of a talk he gave, less than a year before he passed away, at the XXVth International Congress of Psychology in Brussels in the summer of 1992 (when I was recruited as one of the little people in yellow T-shirts, the equivalent of ballkids at Wimbledon).

In one way, Donald Broadbent brought me to William James. In another way, Lyn Hejinian, via the great modernist poet Gertrude Stein (sometimes simplistically but not entirely ineffectively designated "the Picasso of poetry"), brought me to William James. My double inheritance revolves around the one question that has occupied much of my thinking in any domain: How do we choose what we see? Or what we hear, or notice, or perceive, or...Do we have a choice in the matter, or has the choice been made for us by some kind of inflexible cognitive apparatus? The one question soon breaks up in many slightly different versions in the Garden of Forking Questions, to corrupt the title of a short story by Jorge Luis Borges. *The Anatomy of Bias* presents my account of seventeen years' gardening.

Many people have helped me, encouraged me, mentored me, or challenged me. I would like to thank my colleagues, friends, teachers, and students, Arnoud van Adrichem, Hugo Bousset, Marc Brysbaert, Joseph Bulbulia, Roger Carpenter, Tom Carr, Brian Coe, Doug Davidson, Peter De Graef, Dirk Derom, Arkadii Dragomoshchenko, Géry d'Ydewalle, Fernanda Ferreira, Wim Fias, Hiroki Fujiwara, Dave Harper, John Henderson, Andrew Hollingworth, Yoshinori Ide, Shoichiro Iwakiri, Todd Jones, Reiko Kawagoe, Shunsuke Kobayashi, Masashi Koizumi, Jan Lammertyn, Bill Manhire, Kae Nakamura, Kiwao Nomura, Greg O'Brien, Michael Palmer, Masamichi Sakagami, Sue Schenk, Gerd Segers, Muneyoshi Takahashi, Yoriko Takikawa, Jan Theeuwes, Minoru Tsukada, Ken-Ichiro Tsutsui, Johan Velter, Leo Vroman, Johan Wagemans, Katsumi Watanabe, Matt Weaver, Jeff Wickens, Carolyn Wilshire, Regan Wisnewski, and Bob Wurtz.

Most of all, with this book I would like to pay tribute to the one teacher who taught me more than any other, and it will be obvious that that must be Okihide Hikosaka.

The final words of gratitude, from a helpless heart and a human brain, belong to the mysterious world of the unsayable, yet they wish to be said, and I wish to say them here, for Shizuka, Nanami, and Shinsei.

The Anatomy of Bias

1 Bayes and Bias

> Eyes talked in-
> to blindness.
> Their—"a
> riddle, what is pure-
> ly arisen"—, their
> memory of
> floating Hölderlintowers, gull-
> enswirled.

The words of the poet serve as the oracle, the omen or portent of things to come. The strange formulations create a peculiar anticipatory image that colors, biases, or otherwise influences the perception of objects and events. Here, the words are from the poem "Tübingen, January" by the great German-language poet Paul Celan (2001, p. 159), whose work was inextricably linked with the trauma of the Holocaust and achieved what Theodor Adorno had initially thought to be impossible—to make (some kind of) sense after Auschwitz. Of course, it is not easy to explain exactly how the notoriously difficult poetry of Paul Celan *makes sense*, but I think I will try to do just that, be it obliquely, and not right away.

What Good Is a Mystery?

Noisy or nebulous forms of communication will normally receive little sympathy from scientists and other lucid minds who wish to obtain clear information about the world around them. Yet, such clarity must itself be seen as a poetic construction, or an object of wishful thinking, given the level of complexity and the number of stochastic processes involved in our world. Sometimes the best way to approach things in words is simply to start speaking, mark the perimeter, and try to systematically move closer. Even Ludwig Wittgenstein, in his *Tractatus Logico-Philosophicus*, appeared to do something of the sort, following his ominous preface with the stern dictum (Wittgenstein, 2001, p. 3) about when to speak (only when you can do so clearly) and when to be silent (in all other cases, i.e., forever?).

Wittgenstein happily ignored his own dictum with arguably the most mysterious, opaque, and poorly understood philosophical treatise of the twentieth century. He even admitted in the first sentence of the very same preface (p. 3) that "[p]erhaps this book will be understood only by someone who has himself already had the thoughts." The question remains whether anyone has, but that does not diminish the beauty and the attraction of the book, much like the incomprehensibility of Celan's poetry does not stand in the way of the pleasure of reading.

If it was Wittgenstein's intention to write the *Tractatus* in accordance with his dictum about speaking clearly, then his concept of clarity might have more to do with abundance of light and sharpness of contours than with the amount of activity in Wernicke's area, the brain's putative center for linguistic comprehension. This kind of speech, then, would truly represent objects or trains of thought, in a crystallized form, which we can marvel at, preserve, repeat, and return to, time and again. Is it only me, or does this sound like a definition of poetry?

The thing expressed in words would be given a special place, in the spotlight, clearly visible for anyone to see, like a strange prehistoric artifact in the British Museum. The visitor might have no clue what the artifact was used for or what it represented for whoever made it, but that bit of mystery is certainly compatible with a sense of enjoyment in contemplation. Indeed, associations with the abundance of light and the lack of short and easy answers bring us to some of the loveliest entries in the dictionary (e.g., "brilliant," "radiant," "amazing," and "wonderful"). With Gertrude Stein we can further offer the proposition that "if you enjoy it you understand it" (Stein, 2008, p. 10). Turning our attention to the scientific enterprise, we might point out that without an initial mystery there would be no hope for a happy end in the form of a correct theoretical model. Without the excitement of weird questions there would be no thrilling search for answers.

So I will charge ahead, and insert a few lines from some of my favorite poems in a scientific discourse, with the double purpose of prickling the senses and showing how the mere fact of having the senses prickled shapes the subsequent search. The borrowed elements of poetry might generate curiosity, arouse the mind, and heighten the acuity of cognitive processing. The occasional verse would serve as a target or template to guide the conceptual search, exerting top–down control over the way in which incoming signals (the actual scientific data) are selected, integrated, and abstracted. If my plan works, it will serve its purpose as a rhetorical device, echoing the object of study. If it does not, the reader will still have read a few lines of poetry, and that will never bring sadness in my book.

But what, the verse-phobic scientist might wish to know, is the real purpose of your average oracle? In Greek mythology the hero or heroine who seeks advice, and instead receives only garbled instructions, invariably fails to benefit from the trip to Delphi. Worse, it often seems as though the very content of the oracle leads to the protagonist's doom, as when Oedipus, abandoned at birth, ends up killing a stranger and marrying the widow (his true father and mother, respectively). As a narrative technique, though, it works beautifully; it determines the architecture of the story, creating an expectation, opening up a mental slot that needs to be filled. The reader or listener knows what is coming and will not consider the story complete until Oedipus realizes the full horror of his fate.

Making a giant leap, a wild generalization, the complex structure with a prime (an oracle) and a response (the actual fate) looks like a prototypical example of the kind of thought process that some of this planet's most famous researchers (Fitch & Hauser, 2004; Hauser, Chomsky, & Fitch, 2002) have argued to be characteristically human, involving more than mere concatenation of elements or phrases, but hierarchical relations and long-distance dependency, as in an IF...—THEN... structure, in which you can keep embedding other conditions, in principle ad infinitum, but in practice only until your hard drive crashes or your brain forgets what you were talking about. (If, when reading the previous sentence, you get lost, then either I am a bad writer or you may not have what it takes to be human.)

In plain English: the oracle sets up an expectation that helps the reader or listener to interpret the story. Expectations that guide interpretation are an extremely powerful cognitive tool and may be the secret of our species' success. They imply the usage of some, however rudimentary or sophisticated, computational model that generates expectations on the basis of prior information. The process of interpretation can then provide feedback. With new information coming in, expectations are confirmed or disconfirmed, proving or disproving the validity of the computational model. Now, my language must start sounding familiar to readers versed in probabilistic approaches to the study of brain function (for excellent introductions to these, see Doya et al., 2007, and Rao, Olshausen, & Lewicki, 2002).

Expectations that guide interpretation are so basic as to pervade every domain of human thought. At the very heart of science there is the idea, perhaps most elaborately formulated by Popper (2002a), that there should be a kind of logic that underscores discovery, working from theory to hypothesis to prediction, or from general to particular, until we arrive at a well-defined forecast, a concrete expectation, which can be tested empirically. If the prediction comes out, things remain as they are, and we may be happy, but we don't really learn anything; if the prediction proves to be false, we have to go back to the drawing board.

Though this logic for learning is the hallmark of science, similar cognitive sequences operate more implicitly in all our daily activities from choosing where to have lunch to deciding when would be the best time to wade through our e-mail. We constantly juggle beliefs and expectations—about which restaurant serves what quality of food and gets exactly how crowded or about when we can treat fifteen minutes as spare time because we can't really use it meaningfully in any other way. We may develop routines, have preferred restaurants, and tend to check e-mail when we arrive at the lab in the morning, suffering from a brief bout of cognitive inertia after an intense (mostly uphill) twenty-two-minute mountain bike ride. And sometimes we update our list of preferred restaurants, when, during the last visit to our former top choice, the waiter was rude, and the salad, for which we waited a full seventeen minutes, had a fruit fly in it. Or we stop checking e-mail in the morning, as it makes the cognitive inertia even worse, and instead we look briefly at the BBC News Web site, which bores us soon enough, so that we can move on to the real work in the lab with a properly blank and ready mind.

The principal advantage of using mental models that make predictions about the world must somehow resonate with Sir Francis Bacon's "*sciencia potentia est*"—knowledge equaling power—or, more fully, with aphorism III from the first book of *The New Organon*, first published in 1620 (Bacon, 2000, p. 33):

Human knowledge and human power come to the same thing, because ignorance of cause frustrates effect. For nature is conquered only by obedience; and that which in thought is a cause, is like a rule in practice.

Bacon might have been borrowing from the bible's *Proverbs*, particularly 24:5. In the Authorized King James Version, from around the time Bacon was writing in Latin, it reads, "A wise man is strong; yea, a man of knowledge increaseth strength" (Anonymous, 2005). Biblical or not, there does seem to be a deep connection between knowing what will happen when and where, and the opportunity, if not the ability, to do something about it. Individuals, groups, and companies, and occasionally even the United Nations, benefit from devising strategies and aiming for fast and effective action that yields something good or prevents something bad.

It may be wise to be wise, even if this proposition looks suspiciously circular or in danger of infinite regress, with an endless series of "Why?" questions and a relentlessly growing sequence of embedded wisdoms ("Because it is wise to be wise to be wise to be…"). That it is better to know than not to know has generally been taken for granted ever since, or despite, the fall from Paradise (in the parlance of an Abrahamic religion). Clear though its benefits are, the process of knowing itself defies explanation, as if there rests a taboo on understanding our understanding. We may roughly say that it has something to do with getting access to fundamental laws of nature, but how we actually achieve this, and whether (or which of) those laws really exist, remains a matter of debate among the brightest minds of our species, and some of these are very skeptical about whether we will ever understand our understanding (one of the most forceful arguments being that by Penrose, 1989, with a wonderful twist on Gödel's theorem).

One aspect of knowing that has intrigued me personally is how personal the act of knowing often seems to be. Knowing feels more like a feeling than the rational processing of an undeniable truth. Along with understanding and remembering, knowing may have a component of belief in it or an acceptance that some mental images are good enough without questioning them any further, regardless of how they relate to any kind of actual scene, present or past. If I am right, we would tend to be pragmatic and rather minimalist when it comes to the amount of computational power we employ for any particular act of thinking, knowing, believing, and so forth. I see the human mind as a minimalist theorist, or a lazy thinker. By default, the mind would choose the theory of least resistance, or the cheapest concept (hence the early emergence of gods in every known human culture, easy theories as they are, perfectly to blame for everything). Rethinking and changes to the mental model of the world are inspired mainly by adverse runs when things do not go as expected. Simple models gradually get replaced by more complex models as a function of the amount of stress experienced. The level of accuracy required is dictated by the performance of the mental model; if all goes well with a simple but inaccurate model (e.g., Newtonian physics), we might as well keep relying on it.

The central point of this little excursion is that the use of knowledge is subjective, situational, dependent on the actual context in which an individual, a group, a company, or occasionally even the United Nations finds itself. "*Cogito ergo sum*," I would like to repeat after Descartes, but with emphasis on the subjectivity of being, pointing to the fact that both the phrase "I think" and the phrase "I am" have something in the subject slot, even if it is only implied in Latin. With subjectivity comes perspective and limitation in time and in place. To deny the inherent subjectivity is to fall prey to the potentially damaging effects of distortion and bias. Only in the explicit acknowledgement of our own subjectivity, in the willingness to converse with and learn from people who have other viewpoints, can we hope to reach a more objective stance in which all the various idiosyncrasies in thought and feeling are given their due.

The counsel may sound obvious enough, or even slightly naive in its unchecked idealism. Yet how come the obviousness does not obviously translate to practical application? "Seeing is believing," the saying goes, as if the simple act of seeing were the best method of knowledge acquisition, in full denial of its limited validity as a truth procedure. When do we really perceive things, and when do we simply take to be true what fits our understanding of the way things are supposed to go? The process of apprehension with the mind is certainly one in which the quality of data analysis is too often overestimated. Here, I would like to turn to the opening words by Paul Celan. "Eyes" are "talked into blindness," he observed in his characteristically sparing use of words. It sounds disparaging, like a complaint addressed to those who see, but don't really *see*, believing the word (the prediction) rather than what is actually there to see.

How do we move on from the dilemma? If the acts of the mind are inherently subjective, in the sense that they are shaped by the semantic system of an individual, then how can we acknowledge this and work toward a truly objective view of things? Clearly, it will not be helpful to exaggerate the role of the subject. There is little be gained from a caricature view, be it postmodern, constructionist, or simply absurd, according to which there is no such thing as the truth or a knowable actual state of things out there. Somewhere in the middle between the denial of subjectivity and the refutation of reality, we can walk with Celan's poem, in which the "eyes talked into blindness" desire to resolve the riddle of "what is purely arisen." The eyes seem to be aware of their fallibility, and they wish to separate the wheat from the chaff. The peculiarities of our visions, "of floating Hölderlintowers, gull-enswirled," are too concrete and too arbitrary to be mere figments of the imagination. Somehow the mind must have bundled together bits and pieces of reality into a proposition about, or image of, some aspect of the world. The task is to understand how the subject's being in time and space restricts and sometimes skews the information available for processing.

For Paul Celan, this task was all the more urgent as he struggled with his chronic mental illness, a bipolar disorder of a rather malignant nature, with several violent outbursts over the years. Did he see things truly, or was he crazy? The question would have haunted his mind. The end, unfortunately, was tragic; the poet jumped to his death in the river Seine in April 1970 (he had lived most of his adult life as an expat in Paris). Even in the early 1960s, when Celan wrote "Tübingen, January," he must have fully realized what was happening to him; his reference to Friedrich Hölderlin can count as exhibit A (Hölderlin is the archetypal "mad poet," who spent half of his sad life as a recluse in the attic of a mill, not too far from the center of the idyllic and well-preserved town in southern Germany).

The poem only gets starker in the remainder, not quoted here for fear that I would get stuck for another ten pages or so. I will simply note that Celan talks about drowning, the plunging of words, and how a visionary would only babble incomprehensibly "if he spoke of this time" (Celan, 2001, p. 159). It seems entirely possible to me that Celan, in a bout of obsessive–compulsive ideation, ended up believing in the necessary truth of the poem's prediction. If so, this would make it arguably one of the most miserable poems in recent history, as it hung for almost a decade over the patient's head like the sword of Damocles, until Celan finally gave in and committed suicide.

The Role of the Prior

A sword above your head, dangling from a single hair, certainly should give you a vivid impression of imminent danger, and in the case of Damocles it effectively ruined his appetite, as if the sense of disaster in the making (even though the hair in question was good, strong horsehair) prevented his gustatory system from processing the riches that Dionysius II of Syracuse, a proper tyrant, had so generously offered him a taste of. Here, the anticipatory image did more than simply guide the interpretation; it fully dominated the experience and changed the course of action—as soon as Damocles noticed the lethal weapon on a virtual course to pierce his skull, he quickly asked his master if he could be excused from the table.

The legend can be read as a beautiful little parable of the interdependent dynamics of bias, sensitivity, and decision making, or how the fear for one thing dampens the perception of another and elicits an adaptive response, an instance of operant avoidance behavior. At some point in time, though, we would wish to move from parable to paradigm and develop ways to study these dynamics systematically in the formal language of science.

As it turns out, history has already shown that the influences of expectations and subjectivity in perception and cognition can indeed be studied to a surprisingly detailed degree thanks to computational tools based on probability theory. Statistics and the various techniques of likelihood estimation form the perfect platform for the investigation of how we perceive things, think about them, and make decisions. Of course, statistics is crucial to most scientific endeavors, as it formalizes the processing of empirical data, but for the cognitive sciences, from psychology to computational neuroscience, and from neurophysiology to artificial intelligence, statistics takes an even more central position, being a method as well as a metaphor. The myriad acts of mind and events in the brain are all about processing data, and it may not be a crazy thought to think that what happens inside our skull is itself governed by a kind of applied statistics, with data archived in frequency distributions, and hypotheses accepted or rejected on the basis of the available evidence.

The point of departure in statistics and probability theory must be, for now and forever, the work of Thomas Bayes, and particularly his theorem—or rule, if you prefer—about how the likelihood of a particular something is weighted by its prior probability. The prior probability, or simply "the prior," is where subjectivity comes in, where knowledge, expectation, and beliefs can play their part. Bayes's theorem puts the role of the prior firmly in a formula, and its ramifications are the core focus of current psychophysical and neural models of decision making. However, before I turn to these, I would like to taste a sample of the original (Bayes, 1763, p. 4): "If a person has an expectation depending on the happening of an event, the probability of the event is to the probability of its failure as his loss if it fails to his gain if it happens." These are the posthumous words of an eighteenth-century Presbyterian priest, but they sound almost contemporary and are likely to make some kind of sense even to readers who are generally predisposed to get tired quickly from all things mathematical. Here, Bayes makes the straightforward proposal that the numerical data of microeconomics, in terms of likelihood of gain or loss, directly follow the statistics of the real world, in terms of likelihood of events happening or not. The statement may appear somewhat trivial, achieving nothing more than a mere duplication, creating a double, or a new representation of the original. On second thought, however, we may recognize the mechanism as one that enables the creation of a virtual model, a little "toy model of the universe," which would map the physical onto the mathematical, or the ways of the world onto circuits of the brain. Suddenly, this duplication project looks anything but trivial, rather impossibly difficult, a rational ideal. In between the lines we might read a task for the scientist, in comparing how an individual's virtual model deviates from the rational ideal. Any systematic deviation could reveal something peculiar about what it is like to be human, or what it is like to be a particular human at a particular set of coordinates in space-time. Invigorated by the undeniable ambition of this scientific project, we may wish to sample some more from the original Bayes, first in print in 1763 (p. 5):

Suppose a person has an expectation of receiving N, depending on an event the probability of which is P/N. Then (by definition 5) the value of his expectation is P, and therefore if the event fail, he loses that which in value is P; and if it happens he receives N, but his expectation ceases. His gain is therefore N – P. Likewise since the probability of the event is P/N, that of its failure (by corollary prop. 1) is (N – P)/N. But (N – P)/N is to P/N as P is to N – P, i.e. the probability of the event is to the probability of it's [sic] failure, as his loss if it fails to his gain if it happens.

One thing I know for sure is that my own little virtual model cannot cope with this language, and so this is perhaps a good place to admit that I am one of those readers who are generally predisposed to get tired quickly from all things mathematical. My Great Step, or the Problematic Idea of This Book, of necessity will rely only on the most rudimentary type of equations and formulas—the type that are easy enough to capture in words or visual schematic representations. Where relevant, I will add references to the "real deal," papers and monographs with hard-core stuff aplenty for the reader with an insatiable appetite. In the meantime, I am afraid that difficult passages such as the one by Bayes above tend to literally drive me to distraction. I get derailed by surface features; here, I note the odd spelling, "its failure" and "it's failure," inconsistent, incorrect, and sloppy, and so why should I trust the incomprehensible argument?

Even the true masters of statistics, including such luminaries as Ronald Fisher and Karl Pearson, had trouble understanding the ramifications of the original proposal, suggested Stephen Stigler (1982), and the very same skeptic went on to dispute the conventional view that it was really Bayes's proposal in the first place. In a delightful little article for *The American Statistician*, Stigler (1983) upheld his Law of Eponymy, claiming that no discovery or invention is named after whoever really did the work; instead the first person who fails to give proper due would tend to scoop the honor. The piece reads like a true whodunit, with a plausible unsung hero emerging—Nicholas Saunderson, the famous and incredibly talented blind professor of mathematics at Cambridge. This also obliquely raises the question of who really discovered the Law of Eponymy (or who it was that Stigler failed to credit), and to avoid similar misconduct on my part, I will quickly admit having first read about Stigler's doubts in Gerd Gigerenzer's (2002) equally delightful *Reckoning with Risk*.

Back from the past, with our feet firmly on the ground, we will do wise to concentrate on the common understanding of Bayes's theorem in our time. Intuitively, the theorem simply says that likelihood is weighted by prior probability. The end result is a posterior probability, something like our best guess, given all the evidence. "The evidence," then, consists of an actual observation, which constrains the likelihood of a particular hypothesis, in combination with a constant (or normalizing denominator) and the a priori likelihood of the hypothesis. Perhaps a concrete example is in order.

Consider your twelve-year-old daughter (or, if you know of no such instance, consider the twelve-year-old daughter of your neighbor, or of your neighbor's neighbor, or...; the recursive process should be continued until you can think of a proper instance in your neighborhood). Have she and a boy in her class been kissing? Bayes's theorem can tell you how likely it is that this has occurred, given that she blushes (this is the posterior probability), on the basis of three sources of information: the generative model, the prior probability, and the marginal probability. The generative model says how likely an observation is (that she blushes), given that a particular hypothesis is true (that she and a boy in her class have been kissing). Note that this conditional probability, P(Blushing|Kissed), is the exact mirror image of the posterior probability, P(Kissed|Blushing). The prior probability gives the base rate of how likely it is that the (your) daughter and the boy in her class have been kissing, P(Kissed), whereas the marginal probability says generally how likely she is to blush, P(Blushing).

Obviously, if the twelve-year-old girl in question tends to blush quite often in general, but not necessarily when she has been kissing a boy in her class, then blushing tells you much less than if she rarely blushes, yet is sure to blush when she has been kissing a boy in her class. Bayes's theorem captures this mutual dependency of different types of probability, stating that P(Kissed|Blushing) equals the product of P(Blushing|Kissed) and P(Kissed) divided by P(Blushing). Let's say that that twelve-year-old daughter (of yours) generally tends to kiss a boy in her class with a likelihood of about one in twenty, [P(Kissed) = 0.05]. Now if she tends to blushes quite often, [P(Blushing) = 0.4], but not necessarily when she has been kissing a boy in her class, [P(Blushing|Kissed) = 0.7], the posterior probability equals 0.7 times 0.05 (i.e., 0.035), divided by 0.4, or precisely 0.0875. She happens to be blushing? This does not allow you to conclude that she has been kissing a boy in her class. The probability that she blushes, given a kiss of that kind, is less than one in ten. But if she rarely blushes, [P(Blushing) = 0.1], yet is very likely to blush when she has been kissing a boy in her class [P(Blushing|Kissed) = 0.9], the posterior probability equals 0.9 times 0.05 (i.e., 0.045), divided by 0.1, or precisely 0.45. She happens to be blushing? Chances are close to one in two that she has been kissing a boy in her class.

How can we be sure the numbers are correct? Do we simply take the theorem for granted, or can we work out some kind of proof in our own mathphobic way? I was one of the worst performers in math during high school, but I am always willing to try something without too much effort or too many fancy tricks. First, looking at the theorem, we see just four terms, two of which are each other's counterparts. On the left of the equation we have the posterior probability, P(X|Y), and on the right we have its counterpart, P(Y|X), combined with two other terms, P(X) and P(Y). To phrase it exactly: P(X|Y) equals a fraction, which consists of a numerator determined by the product of P(Y|X) and P(X), and a denominator given by P(Y). Let us call this proposition 1, the theorem of Bayes. As a formula it should not look too ominous; all we need to do is multiply one thing by another and then divide the result by something else. That twelve-year-old daughter (of yours) can do it, so you can too.

Of course, the X and Y are just abstract placeholders, so we could easily rephrase proposition 1 by swapping them around, putting an X wherever we find a Y, and vice versa. Proposition 2 then reads as follows: P(Y|X) equals a fraction, which consists of a numerator determined by the product of P(X|Y) and P(Y) and a denominator given by P(X). Now, the funny thing is that we can plug proposition 2 into proposition 1, to remove one of the four terms and replace it with a new combination of the remaining three, each of which is now used twice in the formula left standing. Thus, we do a Bayes on Bayes, apply the rule in the rule, or rely a little on our beloved mechanism of recursion in the hope of proving the whole.

Having done the deed, we have this: P(X|Y) equals a fraction, which consists of a numerator determined by the product of *the right side of the equation in proposition 2* and P(X) and a denominator given by P(Y). This actually puts a fraction inside a fraction, making the formula look scarier if you use conventional notation than if you say it in words. Anyway, to simplify things, we can use the multiplication principle to carry P(X) and P(Y) to the left of the equation, leaving only *the right side of the equation in proposition 2* on the right. As for the left side, P(X) will end up being the denominator, whereas P(Y) will join up with P(X|Y) to form the product that defines the numerator. The multiplication principle and the rules about carrying terms across to the other side of the equation were properly etched in my memory. I will take the liberty of assuming that a similar kind of etching would have occurred for anyone picking up a book like the one before us.

Thus, we have a fraction on either side of the equation. On the left we have P(X|Y) times P(Y), to be divided by P(X). And on the right we have the right side of the equation in proposition 2, which claimed that P(X|Y) times P(Y) should be the numerator, and P(X) the denominator. The left says exactly the same as the right, quod erat demonstrandum!

Or not? Did I create nothing more than a tautology, showing that the rule is true if the rule is true? What would happen if you swap P(X) and P(Y) in proposition 1 and then apply my recursive trick? Again the tautology works, saying that the new rule is true if the new rule is true, which it probably is not.

Well, at least we have acquired some practice in playing with the terms, and other than that, I would like to draw three conclusions from this little exercise: (1) It is fun to think for ourselves, we should do it more often; (2) recursion is not the answer to everything; and (3) we had better move on and stick with the original motto, saying this is not a book of mathematics. An unflappable skeptic might point out that conclusions 1 and 3 are somewhat contradictory, but I guess in this matter the limits of time and the reader's patience should prevail.

The proper proof of Bayes's theorem is actually quite straightforward, I must admit, though I did not manage to come up with it myself (for a more serious primer, see Doya & Ishii, 2007). All we need is a little detour via the definition of conditional probability, or the probability that one thing is true given that another is true, $P(X|Y)$, that is, the type of term we are already familiar with from Bayes's theorem. The definition of conditional probability is based on another, in fact more basic, concept: joint probability, or the probability that two things are both true at once, P(X is true *and* Y is true), sometimes notated as $P(X \cap Y)$.

The definition of conditional probability, then, states that the probability of X, given Y, equals the joint probability of X and Y divided by the general probability of Y. That is, $P(X|Y) = P(X \cap Y)/P(Y)$. This naturally makes sense if you think about it. We might want to follow Gigerenzer's (2002) advice, and reason in natural frequencies for a minute. Take all the cases in which the twelve-year-old girl had been kissing a boy in her class; in how many of those cases did she also blush? This number, how many times blushing and kissing out of how many times kissing in general, basically gives you the desired conditional probability.

A good statistician should like to play around with any definition. So did Bayes, one of the very first of that species of human. If it is true that $P(X|Y) = P(X \cap Y)/P(Y)$, then we can also move $P(Y)$ to the other side of the equation to give us $P(X \cap Y) = P(X|Y) P(Y)$. Defining the opposite conditional probability, $P(Y|X)$, we get $P(Y|X) = P(X \cap Y)/P(X)$, which we can rewrite as $P(X \cap Y) = P(Y|X)P(X)$. So we are basically rewriting $P(X \cap Y)$ in two ways:

$P(X|Y) P(Y) = P(X \cap Y) = P(Y|X) P(X)$.

Now we can get rid of $P(X \cap Y)$, saying $P(X|Y)P(Y) = P(Y|X)P(X)$. To get his theorem, Bayes then only had to move $P(Y)$ to the right of the equation: $P(X|Y) = P(Y|X)P(X)/P(Y)$. Quod erat demonstrandum! (For real, this time.)

By now I have spent much more time talking about, and circling around, Bayes's theorem than would have been needed to simply restate it in textbook format. Even if you were not schooled as a neuroscientist, you will know it by heart. Hopefully, you will also have developed some intuition about decision making as an inherently statistical enterprise, even if we shy away from numbers and formulas when weighing options and taking things to be true or not as we go about our business and do our mundane doings in daily life. However, Bayes's theorem says something more specific than that.

The main point to take away from the Bayesian way of looking at the world is this: Our beliefs about the world should be updated by combining new evidence with what we believed before, "the prior." The role of the prior is to color, or to help interpret, new information. Does blushing mean that the twelve-year-old girl has been kissing a boy in her class? Our prior beliefs about her kissing behavior and blushing tendencies will help us draw better conclusions than we would reach if we were to consider only the current evidence. Decision making stands to benefit from keeping track of how often things happen in the real world. Even the most rudimentary records and nonparametric statistics (relying only on rank ordering as in "This happens more often than that") are likely to improve perception, categorization, and all the more complex forms of cognitive processing to define things as they are. The study of decision making translates into the problem of uncovering exactly how an individual, a group, a company, or occasionally even the United Nations makes use of different kinds of information about the likelihood of things and events. The role of the prior, here, determines the potentially idiosyncratic characteristics of how decisions might tend to go one way rather than another. The prior determines bias.

I have dropped the Heavy Word. Bias, what does it mean exactly? According to *A Dictionary of Statistical Terms* by Kendall and Buckland (1957, p. 26), it is as follows:

Generally, an effect which deprives a statistical result of representativeness by systematically distorting it, as distinct from a random error which may distort on any one occasion but balances out on the average.

The statement is a bit terse, as we might expect from an old-school statistical dictionary (picked up practically for free at a sale of unwanted books at the City Library of Wellington), but the gist is clear. Bias implies something systematic, nonrandom, which pulls the statistical result, or decision, in a particular direction, away from the neutral. The loss of neutrality, or the introduction of subjectivity, carries a load of negative connotations in daily life as reflected in the lemma for bias in *The Pocket Oxford Dictionary of Current English* (Thompson, 1996, p. 75):

n. 1 (often foll. by towards, against) predisposition or prejudice. 2 Statistics distortion of a statistical result due to a neglected factor. 3 edge cut obliquely across the weave of a fabric. 4 Sport *a* irregular shape given to a bowl. *b* oblique course this causes it to run.—v. (-s- or -ss-) 1 (esp. as biased adj.) influence (usu. unfairly); prejudice. 2 give a bias to.

Except in weaving or bowling, bias is associated with a number of known villains: prejudice, distortion, and the adverb for lack of fairness. It is only one step shy of racism, sexism, nepotism—a host of—isms that good citizens, if not good politicians, would like to stay away from. Whether I am a good citizen I will leave for others to judge, though on the topic of *The Pocket Oxford Dictionary of Current English* I should note that the copy inexplicably found its way to my personal library (it had belonged to the first author of Shimo & Hikosaka, 2001, a rogue reference offered in compensation for the excessively long duration of borrowing the pocket dictionary, which, by the way, even if I pocketed it, fits in no pocket of mine). Good citizen or not, I think the word "bias" did not get a fair shake. Obviously, the unfair treatment of others represents an ugly disease in human society, one that we should prevent and remediate in any way possible, but to simply equate bias with something bad may be throwing the baby out with the bathwater.

Bias is part and parcel of Bayesian reasoning, emphasizing the crucial role of the prior in the assessment of probabilities, representing beliefs about how one thing might be more likely than another. The prior is exactly the term that modelers of Bayesian inference in perception employ to characterize observer biases (e.g., Mamassian, 2006; Mamassian & Landy, 1998). Such biases are perfectly rational if they correspond with the statistical regularities of the environment. Perhaps we should dust our vocabulary and heed once more the words of our favorite fourteenth-century Buddhist priest from Kyoto (Kenkō, 2001, p. 13):

The same words and subjects that might still be employed today meant something quite different when employed by the poets of ancient times. Their poems are simple and unaffected, and the lovely purity of the form creates a powerful impression.

The word "bias" deserves to be exonerated, polished, and used properly. Instead of dismissing bias as a form of evil, I propose we should acknowledge bias as a fundamental property of human thinking, perceiving, and decision making. If we are to eradicate the social crime of prejudice, we should establish whether and how prejudice derives from bias and whether and how bias derives from statistical regularities in the world. The derivation of prejudice would represent the real evil, the one we would wish to redress. However, perhaps the best way to do so is not by denial but by explicit formulation of bias and the extent to which it is rational. Put differently, we need to learn to distinguish "good bias" from "bad bias." In the meantime, the best way to start the enterprise is by considering the basic role of bias in decision making. For this we need to wrestle with probability distributions.

Before doing so, I would like to pay one final tribute to the lovely and pure, if slightly incomprehensible, ancient words of Thomas Bayes by reciting my all-time favorite title for a scientific monograph (one that Bayes published anonymously in 1736): *An Introduction to the Doctrine of Fluxions, and a Defence of the Mathematicians Against the Objections of the Author of The Analyst*. Bayes's defense was for Newton against Berkeley (the author of *The Analyst*), and though I searched hard for it on the Internet, I found no free online version. I can only imagine what Bayes's argument sounds like, but the word "fluxions" appeals to me, and the defense of mathematics against analysts somehow rings a bell for me, with contemporary neuroscience, and its computational approaches, on the defense against present-day psychoanalysis, represented by post-Lacanian thinkers such as Slavoj Žižek—a battle that may be outside the field of vision for many neuroscientists but is nowhere near dying down in cultural studies, including literary theory, a field that interests me for idiosyncratic reasons. So I, too, wish to introduce fluxions, but then fluxions in the form of movement between computation and metaphorical intuition, not to have one pushing the other out of the ring but to get the best of both worlds if that is at all possible—hence the poetry as well as the wrestling with distributions.

Wrestling with Distributions

One of the first things I learned, to my dismay, when I started collecting real, heavy-duty, hard-core scientific data was how massively variable they were. In my maiden project, a partly tongue-in-cheek exploration of the eye movements of a poetry critic, I was not particularly worried about that, thinking it was a crazy project anyway (Lauwereyns & d'Ydewalle, 1996). But when I then started collecting data for my PhD thesis in a very classic visual search paradigm, I was positively baffled by the fact that so simple a task as pressing a button when you find a letter Q among distractor letters O on the computer screen could lead to such vastly different response times, with some of my victims (first-year students in psychology) taking forever to find the target (more than a second, occasionally even two) and others finding it right away (in less than half a second). Even for the same participant, the data often looked very messy, with response times all over the place, sometimes 300 milliseconds, sometimes 800.

Apparently, people were unable to exactly replicate what they did, though I asked them to do the same thing for hundreds of trials in a single session of less than an hour, keeping all factors constant as best I could: the same participant, the same apparatus, the same task, the same events, the same time of day...It dawned on me, slowly, that concepts such as the variability and the standard deviation were crucial to the science I found myself in. My data looked messy, but no messier than those from other laboratories—I was still able to draw publishable conclusions (Lauwereyns & d'Ydewalle, 1997). Investigating the mechanisms of visual perception and decision making involved wrestling with distributions; there was no escaping it. I remembered a remark made in one of my undergraduate classes about how someone, Francis Galton probably, had once said that variability was a blessing in statistics, more important even than the mean of a distribution.

Rather than remaining petrified in the face of variability, I had to record it, chart it, and make it visible in numbers and graphs. How often does a particular event happen? How frequent is it? How frequent is it relative to other events? Moving from observation to abstraction, I was working with probability distributions before I knew it. Measuring the response times of my participants, I would mindlessly apply the descriptive and inferential statistics that are standard procedure in the research field, computing the means and the standard deviations and performing fancy analysis of variance—a few clicks and button presses and out came a set of results that psychologists of a previous generation would have labored on for hours, days even. I was able to make perfectly sanctioned statements about factors that did or did not have a statistically significant effect on response time, without really understanding how I could say what I was saying. Things changed only when I had to teach the materials to others; I took a good look at the textbooks, practiced a great deal in the labs that I was volunteered to be in charge of, and finally started seeing some light at the end of a dark statistical tunnel.

After a while, you can even develop some kind of aesthetic appreciation for the beauty of distributions. Figure 1.1 is, hopefully, a case in point, showing two sets of three distributions—continuous probability distributions, to be precise, which depict the likelihood of all possible outcomes for a given measure, say, response time in my visual search task or weekly ticket sales for movies. The horizontal dimension gives the possible outcomes—ticket sales from zero to a hundred million dollars. The vertical dimension provides the actual probability associated with each outcome, which must be very low indeed for a hundred million dollars, just once in a blue moon, that is, the opening week of a Batman movie, thriving on the ghostly appearance of an actor who had died of an apparent overdose a half year before the movie's release.

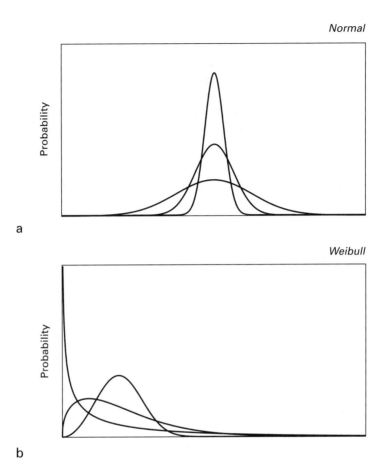

Figure 1.1
Two different types of probability distribution. (a) Three examples of the normal distribution, with the same mean but a different variance. (b) Three examples of the Weibull distribution, with different shapes, morphing from exponential to normal.

The horizontal dimension should contain all possible outcomes, whereas the sum of all probabilities should be one, if we talk in proportions, or a hundred, if we deal in percentages. Let us take a closer look at a few distributions. The three examples in the top panel a of figure 1.1 are instances of the normal distribution, sometimes termed Gaussian in honor of the German mathematician Carl Friedrich Gauss but perhaps more commonly known as "the bell curve." The normal distribution is determined by a location parameter (the mean) and a scale parameter (the variance). The examples in figure 1.1a have the same mean but a different variance—the higher the variance, the flatter the appearance. Note how all three distributions look symmetrical. Perhaps they look too neatly symmetrical? We should be able to think of other shapes for distributions.

It is in fact possible to quantify shape, as shown with the three examples of the Weibull distribution in figure 1.1b. This distribution was named after the Swedish engineer Wallodi Weibull, who specialized in the study of the strength and durability of materials, often facing data that simply could not be fitted into a normal distribution. The Weibull distribution, like the normal, has just two parameters: Again, one is for scale, but the second determines shape instead of location. This makes the Weibull distribution particularly flexible. Depending on how you set the shape parameter, it morphs into an exponential distribution (like the example in figure 1.1b that swings down from top left) or a normal distribution (like the symmetrical example with the rightmost peak), or anything in between (like the asymmetrical example that sits in the middle).

Wrestling with distributions, then, comes down to categorizing events and deciding whether a particular observation belongs to this or that distribution. We can work in two directions. In most cases we will start from a particular situation, or experimental manipulation, and then collect data to compare the distributions in one case against the other. This is the default approach for a scientific experiment, comparing an experimental condition against a neutral or control condition, which ideally is as similar as possible to the experimental condition except with respect to one factor or dimension—the factor under investigation. For instance, we might be interested in the effects of caffeine on visual search performance. We could perform the experiment in several ways, working with the same or different participants in the two conditions, working with different types of visual search task, caffeine solution, dosage, and so on, and we would have to think hard about how to ensure that a host of other things (e.g., placebo effects, practice with the task, fatigue, boredom) do not contaminate our data, but the bottom line is that we would create an experimental condition with caffeine and a control condition without caffeine. Then all we have to do is write down our observations of response times in two distributions and measure to what extent these overlap. If the overlap is complete, we can safely conclude that the caffeine had no effect. If the two distributions show some degree of separation, we can start thinking that the caffeine managed to do something after all, like increasing the speed of visual search performance. Statistics will give us numbers to support decisions about when the data from two conditions show a "significant" difference. The entire rationale for drawing these conclusions is quite complex when spelled out, but the basic idea is simple: We want to avoid mistakes, so we use estimates of how likely we are to make a mistake if we claim that our observed distributions prove the two conditions to produce different results. (Usually statistical software does this for us in the form of p values.) If the likelihood of error is only one in twenty, or even less, common practice says we can go ahead and make claims about there being a difference.

Perhaps a slightly counterintuitive approach is to work in the opposite direction, from observations back to guesses about which distribution they belong to. Yet this is probably a good characterization of what must happen in the brain when decisions are computed on the basis of the available evidence in terms of activity levels of neurons that represent different alternatives. To see how this works, let us consider a neurophysiological experiment in which we record the electrical impulses of, say, a neuron in secondary (or higher order) visual cortex while the subject is presented with visual stimuli. The subject will usually be a cat or a monkey, but occasionally a human, undergoing neurosurgery (e.g., Quiroga et al., 2005), and the visual stimuli could be anything from former presidents of the United States to random groups of dots moving this way or that.

Thus, for example, we pull up Jimmy Carter on the screen and check what the neuron does in response. We might notice that the neuron becomes particularly active, or tends to fire many spikes, whenever it is Jimmy Carter, but not George H. W. Bush. Can we work the other way around?

We could continue running the experiment but now avoid looking at the screen. Some stimuli are being presented, but we have no clue who or in what order. If we listen only to how often the neuron spikes, can we deduce which former president must have been presented on the screen? How many spikes must the neuron fire for us to conclude that it was Jimmy Carter? Metaphorically speaking, these are exactly the types of questions that other neurons in the brain would be faced with when weighing the input they get from neurons in secondary visual cortex.

The logic unfolded is probably the most powerful approach in contemporary neuroscience when one is trying to model the mechanisms and algorithms seen in neural circuits for decision making (see Gold & Shadlen, 2001, for a bright introduction). The approach was pioneered in the 1960s and 1970s by David Green, R. Duncan Luce, and John Swets (e.g., Green, 1964; Luce, 1963; Luce & Green, 1972; Swets, 1961) and found its definitive formulation in *Signal Detection Theory and Psychophysics*, a book published by Green and Swets in 1966, one of the very few unmistakable classics in this area (the reprint in my collection dates from 1988). The original concern seemed to be all about the purity of signal processing:

The approach discussed here clearly isolates the inherent detectability of the signal from certain attitudinal or motivational variables that influence the observer's criteria for judgment.... To the stimulus-oriented psychophysicist, this analysis is a methodological study, but one that is clearly pertinent since it claims to provide an unbiased estimate of what, for the stimulus-oriented psychophysicist, is the major dependent variable. (Green & Swets, 1988, p. 31)

Green and Swets were clearly rooting for the ideal observer, though they quickly realized that their "Theory of Ideal Observers" (chapter 6) provided an excellent opportunity for "Comparison of Ideal and Human Observers" (chapter 7). In later work, it seemed that John Swets in particular became more and more interested in the broad merits of understanding bias rather than developing a bias of his own against the topic of bias (Swets, 1973, 1992).

Let us borrow the concepts of signal detection theory for a visual schematic representation in figures 1.2, 1.3, and 1.4. The best place to start is by considering the simplest possible decision-making task. Going back to our example with the neuron in secondary visual cortex, we can envisage a forced-choice situation in which the owner of the neuron is simply asked to indicate, in a number of trials, whether a target or "signal" is present, "yes" or "no," just two alternatives. We might get our observer to press a button whenever he or she sees Jimmy Carter. If there is no target, the observer should refrain from pressing the button.

Now we can start recording spikes and button presses and try to relate the former to the latter in our search for a neural correlate of perceptual decision making. Signal detection theory is an invaluable tool in this enterprise, as it allows us to distinguish between two basic ways in which decision making can be influenced. Without diving too deep into the algorithmic depths of the theory, I promise we will be able, a few pages from now, to marvel at its principal strength in teasing apart mechanisms of bias (see figure 1.3) and of sensitivity (see figure 1.4). To fully appreciate how these two ways are fundamentally different, but not mutually exclusive, we first need to come to terms with the basic framework.

As the observer (the owner of the neuron under investigation) makes a decision about the presence or absence of Jimmy Carter, there are logically four possible outcomes: (1) a correct rejection, which occurs when the observer, presented with George H. W. Bush, reports there is no signal; (2) a hit, which occurs when the observer correctly reports the presence of a signal; (3) a miss, which occurs when the observer fails to detect Jimmy Carter actually present among the noise; and (4) a false alarm, which occurs when the observer erroneously reports the presence of a signal.

Green and Swets suggested that these four outcomes could be accounted for with a model that incorporates a "noise distribution," a "signal distribution," and a "criterion" (see figure 1.2, with indications of the four possible outcomes). The two distributions can be thought of as one probability distribution broken down in two "subdistributions," one indicating the likelihood of observing a particular number of spikes given the presence of a signal (i.e., signal distribution) and its complement indicating the likelihood of observing a particular number of spikes given the presence of only noise (i.e., noise distribution). How does our observer decide whether seven spikes should be taken as evidence of Jimmy Carter?

The terminology brings Bayes's theorem back to mind, and indeed, working our way inside figure 1.2, we can recognize the different components of the theorem at play. So let us say that we get a reading of seven spikes during a particular trial in our experiment with the observer looking for Jimmy Carter. What is the likelihood that the stimulus is indeed Jimmy Carter, given a reading of seven spikes? To compute P(Carter|Seven), as we have duly learned by heart, we should work out P(Seven|Carter) times P(Carter), divided by P(Seven).

P(Carter) refers to the entire signal distribution and its relation to all possible cases. In our two-choice task, there are only two possibilities: Carter (signal)

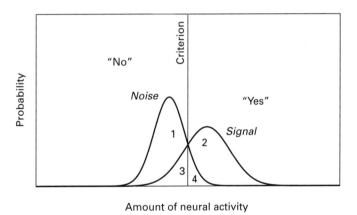

Figure 1.2
An application of signal detection theory. The horizontal dimension represents the number of spikes fired by a neuron. Shown are two hypothetical normal probability distributions, for the neural responses in the case of "noise" versus "signal." The vertical line cutting through the two distributions represents a criterion for signal detection, saying "yes" for spike values above criterion and "no" for those below criterion. There are four possible outcomes: *1* represents the "correct rejections" (saying "no" when there was in fact no signal), *2* shows the "hits" (saying "yes" when there was indeed a signal), *3* points to the area of "misses" (saying "no" although there was actually a signal), and *4* indicates the area of "false alarms" (saying "yes" even if there was really nothing but noise).

or not-Carter (noise). This implies that P(Carter) equals 1 − P(Noise). Thus, we can compare the size of the entire signal distribution to that of the noise distribution—in figure 1.2 they are the same size, that is, we have just as many signal as noise trials in our experiment, or P(Carter) = 0.5. This component represents the prior, and it is easy to see that the basic likelihood of the signal, or our belief of how likely it is, has a large impact on all computations to follow. The numbers would certainly be very different if the experiment had only one Jimmy Carter appearing in every hundred trials.

We can find P(Seven|Carter) by considering the signal distribution to be a complete probability distribution on its own—or multiplying the signal distribution with a factor that brings the total sum of all its cases to one. If P(Carter) = 0.5, we simply need to multiply by two. Now we can trace the curve of the signal distribution until we reach the value of seven on the horizontal axis. The associated value on the vertical axis, multiplied by the appropriate factor, gives us P(Seven|Carter).

To read the general probability of seven, P(Seven), we should not multiply the signal or noise distributions but instead simply trace each of the two distributions until we reach the value of seven on the horizontal axis, read the associated probabilities, and compute the sum of both values.

Now we already have all the components that we need for solving the equation, but in figure 1.2 we might as well look up P(Carter|Seven) more directly, by locating the value of seven on the horizontal axis, then moving up vertically until we hit the curve of the signal distribution, and reading the associated probability. Next we do the same for the noise distribution. Effectively, we find the same two probabilities that we made use of to compute P(Seven). This time we can consider these to make up a total of one, that is, we need to multiply P(Seven) by a factor that brings it to one. Now we can multiply the individual probabilities by the same factor to give us the sought-after number: P(Carter|Seven).

One way or another, the visual scheme presented in figure 1.2 does incorporate the truisms of Bayes's theorem. However, rather than performing these somewhat tedious computations, there is nothing to stop you or me from working more intuitively with the logic and leaving the numerical applications, proofs, and annotations for another day in another life. One thing glaringly absent in Bayes's theorem is an instruction on how to interpret whatever probability we do compute. The theorem might help us wrestle with distributions, but it does not specify what we are to do with the posterior probability once we have computed it. Somehow, we should try to link the posterior probability to a decision or an action. We need a decision rule.

Actually, finding a good decision rule should not be too difficult. A simple adagio would be to try to maximize gain and make sure that our decisions on the presence or the absence of a signal, with the four possible outcomes of hit, miss, correct rejection, and false alarm, combine to our profit. Here, signal detection theory provides us with its most ingenious addition to Bayesian thinking—the concept of a criterion or threshold. In one sense, this is nothing new, merely a formal application of a common practice in statistics, where categorical decisions are imposed on continuous distributions in the form of conventional criteria for "statistical significance." However, in signal detection theory, the criterion is introduced as a borderline that cuts across the signal and noise distributions, enabling one to clearly visualize how the positioning of this borderline determines the likelihood of each of the four possible outcomes.

In figure 1.2 we see that the criterion is taken as the borderline between "yes" and "no" responses. For spike counts higher than criterion, to the right of the borderline, our observer would conclude that Jimmy Carter was shown on the screen. For spike counts below criterion, the answer would be "no." Any case belonging to the portion of the signal distribution to the right of the criterion would then produce a hit (area 2 in the figure), but if the observation of a spike count above criterion actually belonged to the noise distribution, our observer would make a false alarm (area 4 in the figure). Conversely, we can see how this scheme relates misses (area 3) and correct rejections (1) to the positioning of the criterion.

In figure 1.2 the criterion is placed right at the crossroads between the two distributions, at the point where the spike count is equally likely to reflect a signal or noise. To the left of the criterion, the noise distribution dominates, with spike counts that more likely reflect noise than a signal, whereas the signal distribution rules to the right of the criterion. In fact, the criterion is quite strategically (rationally!) placed to minimize the likelihood of an erroneous decision, be it a miss or a false alarm.

Research on eye movement control in macaque monkeys suggests that this kind of categorical threshold idea makes for a plausible neurophysiological mechanism. Doug Hanes and Jeff Schall (1996) showed that the activity of neurons in the frontal eye field (the prefrontal cortical structure for voluntary control of eye movement) consistently peaked at around a hundred spikes per second right before the initiation of an eye movement, regardless of how long it took for the neural firing rate to grow to that peak, and regardless of how long it took for the monkey to initiate the eye movement. When the spike rate was at 100 spikes per second, the eye movement took off. Data from similar experimental paradigms, recorded from neurons in superior colliculus (the major subcortical station that drives eye movement initiation), provided additional support for the existence of an absolute threshold (Krauzlis & Dill, 2002; Paré & Hanes, 2003).

Perhaps the threshold idea is not a crazy one. Applying a rule like that is certainly not difficult and really involves no thinking. All that is needed is enough heat from electrical impulses to wake up the next layer of neurons. It is at about the right level of simplicity to be useful in mapping decision-making properties onto neural circuits. But returning to a safer level of abstraction for the time being, we can explore the effects of positioning the criterion somewhere other than strategically in the middle between noise and signal. Figure 1.3 reproduces the neutral case of figure 1.2 and brings up two other cases for comparison: one in which the criterion is shifted to the right, and another case with a criterion shift in the opposite direction. It is easy to appreciate that the position of the criterion determines the likelihood of different types of errors. With a rightward criterion shift, as in panel b, we avoid false alarms but are much more likely to miss actual signals. In contrast, we reduce the misses at the expense of false alarms if we shift the criterion to the left as in panel c.

When misses and false alarms are equally costly in economical or evaluative terms, the most rational strategy will be to place the decision criterion so that both types of error are minimized as much as possible. In other situations, it may be important to avoid misses—like when we interpret the data from a diagnostic test for pancreatic cancer—whereas false alarms carry less weight. The decision criterion would then better be shifted to the left, minimizing the area of the signal distribution that falls on the wrong side of the criterion, at the expense of an increased number of false alarms. In yet other situations—when we point a rifle at a cloud of dust, kicked up by, potentially, an armed insurgent—it would be crucial to avoid making a false alarm and shooting an innocent victim. Here, the preferred option should be to shift the criterion to the right.

The shifts of criterion install observer biases, leading to different actions or decisions even if the underlying signal and noise distributions retain the same outlook. With rightward shifts, our observer applies a conservative criterion, requiring more evidence than in the neutral case before agreeing that Jimmy Carter is present. On the other hand, more liberally minded observers might shift the criterion to the left and be happy to decide, on the basis of relatively few spikes, that the signal is there all right. The criterion sets the amount of evidence or information required for a decision, and any decision that is predisposed in favor of, or against, accepting a signal can rightfully be called "biased," and yet might still be appropriate, reasonable, or even rational.

I should point out that there are other ways to conceptualize shifts of criterion. My favorite, and a neurophysiologically plausible way, is to shift both distributions, while keeping the threshold in place (Lauwereyns et al., 2002a,

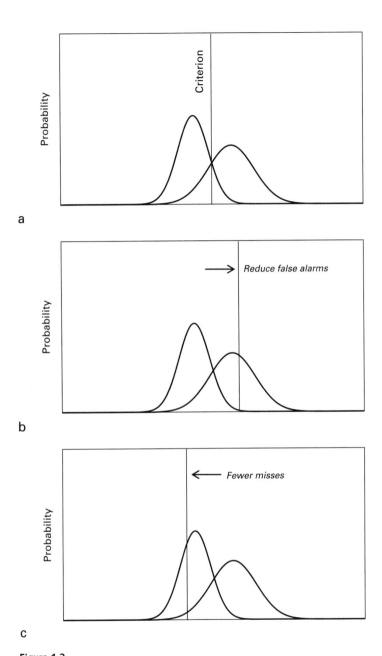

Figure 1.3
Variations to the tune of bias. (a) Take this to be the neutral case, the same as in figure 1.2. (b) The criterion is shifted to the right, more "conservative" than the neutral case, requiring a higher neural firing rate before accepting that there is a signal, and so reducing the likelihood of making a false alarm. (c) The criterion is shifted to the left, more "liberal" than the neutral case, already happy with a lower neural firing rate to say "yes," which brings down the likelihood of missing a signal.

2002b). In practice, this explanation is perfectly interchangeable with the variations to the tune of bias as shown in figure 1.3, and I consider the two proposals to be equivalent. I will introduce the neurophysiological data in detail in chapter 2.

Yet another way to implement bias, definitely a theoretical possibility but as yet not seen in neurons, is to move the criterion to the left or right so that it stays perfectly in the middle between signal and noise. This would be done by more literally applying the role of the prior, that is, by enlarging or reducing the signal distribution relative to the noise distribution (without changing the shapes or the means of either distribution). With an enlarged signal distribution, for instance, the midway crossover point between the two distributions would shift to the left. The enlarged signal distribution could reflect an actual increase in the likelihood of a signal (as when we now present Jimmy Carter on two thirds of the trials in the experiment), or it might reflect an observer's overestimation of the true likelihood—a distorted image of reality. But the conjecture that the shapes and means of the distributions remain the same makes it hard to translate this possibility into a neurophysiologically plausible model of bias in decision making. The real decision making would have to be done outside of the model, with different weights of the signal distribution in the workings of some mysterious Master of Shadows, a decision maker hidden from view. When it comes down to neurons, I would like to see them actually do something if they are to contribute to decision making.

Figure 1.4 shows an entirely different mechanism influencing decision making. Here, the movements and variations occur to the tune of sensitivity. Again we start from the neutral case, the one introduced in figure 1.2. The task of decision making is particularly challenged by the overlap between signal and noise distributions. The overlap implies uncertainty and increases the likelihood of error. Arguably the ideal way to improve decision making, then, would be to try to reduce the overlap, or improve the signal-to-noise ratio so that the two distributions are more clearly distinguished. Assuming that each of the two distributions has a normal shape, we could heighten the sensitivity for a signal by fine-tuning so that both distributions have a crisper appearance with smaller standard deviations, as shown in figure 1.4, panel b. Alternatively, the signal-to-noise ratio can be improved by moving the two distributions further apart, changing the means but not the standard deviations, as shown in panel c. In both cases, we can easily place the criterion at an optimal spike level that succeeds nicely in segregating signals from noise.

To effectively improve the Jimmy Carter-to-noise ratio, we might finally allow our observer to put his or her glasses on. Or we could dim the lights in the room so that the screen stands out. Real-life examples of improved

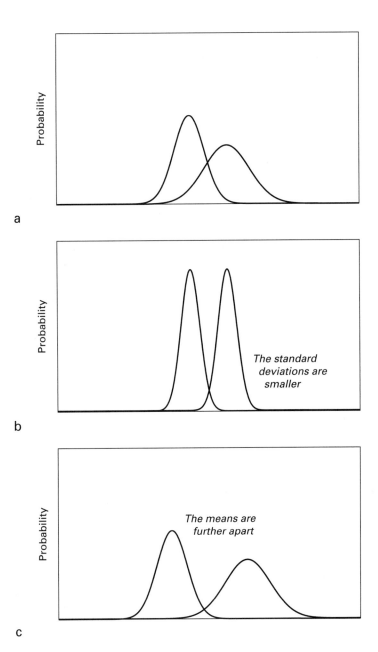

Figure 1.4
Variations to the tune of sensitivity. (a) We start from the same position as in figure 1.2 and 1.3a. (b) Here, the overlap between the two distributions is reduced, facilitating the extraction of signals from noise by decreasing the variance (giving smaller standard deviations). (c) In this case the overlap is reduced by changing the location parameter, leaving the scale parameter untouched. That is, now the means of the two distributions are further apart, whereas the standard deviations remain as they were.

sensitivity often reflect some kind of physical change, creating better conditions for signal reception, but the enhanced decision making can sometimes also be achieved by internal, cognitive operations like paying attention, thinking twice about the data, or double-checking the numbers.

In any case, with whatever degree of bias or sensitivity, the conceptualization of decision making with signal detection theory quickly raises many questions, about the shapes of the distributions, the nature of the decision rules, and so forth, but the framework has the considerable merits of simplicity, specificity, and testability. We will be able to derive predictions from it in terms of neural signatures. One limitation of the signal detection theory, however, is that the logic with categorical decisions is difficult to convert into predictions about response time during decision making. Yet, response times are just possibly the most powerful behavioral measure of what is going on in the brain. They might tell us more than only the accuracy of "yes" or "no," or the trade-off between response speed and decision quality. Time could well provide a quantitative measure of the decision process, of how hard it was, how much thinking it took, and how many neurons had to have their say. Can we work toward some kind of integration of signal detection theory and response time measurement?

Time and the Measurement of Mind

The desire for a quantitative approach to the study of brain and behavior has surfaced only recently, but perhaps it is fair to say that the wishful thinking had been there for centuries, at least since the amazingly modern proposals of Doctor Mirabilis, Roger Bacon—possibly, and horribly, better known as the thirteenth-century model for Sean Connery in the film version of *The Name of the Rose* (see Cregg, 2003, for a preferable biography). To really make progress with numbers, though, and to stimulate the desire further, humans first needed to invent reliable clocks that could tick away the seconds, and then, toying in the lab, stumble on the brilliant idea that these clocks could be useful for the as yet unnamed venture of neuroscience. As it happened, the original proposal emerged in my native language, Dutch, exactly a hundred years before I was born—good enough reason, I would like to think, for a fetishist attachment. Here it is, radically unfiltered and incomprehensible to most (Donders, 1869, p. 119):

Maar is dan ten opzichte der psychische processen iedere quantitatieve behandeling uitgesloten? Geenszins! Een gewichtige factor scheen voor meting vatbaar: ik bedoel den tijd, die tot eenvoudige psychische processen wordt gevorderd.

The idea that we could measure the mind in seconds is certainly an outlandish one, not necessarily understood any better if it is formulated in English. W. G. Koster made a complete translation for the second volume of *Attention and Performance* that he edited in the year that I, of course, do not associate with any landing on the moon, it being a hundred years after a hundred years before I was born (Donders, 1969). Reasoning with time is not easy, but Franciscus Cornelis Donders suggested we should try. Here is my own translation of the excerpt:

> But is then every quantitative approach impossible with respect to mental processes? Not at all! An important factor seemed amenable to measurement: I mean the time taken up for simple mental processes.

In French, we would have spent thousands of inebriating pages *In Search of Lost Time*, but this is the best of Dutch, fully exhibiting its pragmatic quality. Even if the theory was wanting, and the rationale idiosyncratic or simply absent, the intuition that it might be useful was all the incentive required to commence with experiments. Thus, Donders went on to develop his infamous subtraction method, still a standard tool today, comparing the response times of his subjects as they performed different tasks, with systematic variation of the level of complexity for stimulus processing and response preparation. The more complex the mental process, the more time it took to give a correct response.

Somehow the differences in response times did, in fact, correspond with the complexity of cognitive operations, and one way or another, the fact that thinking took time had to be an important observation. For one thing, it suggested that the mechanisms of thought left a material trace, one that was not easily reconciled with Cartesian dualism and its profound divide between the immaterial world of the mind and the physical reality of the body. However, even if the hard-core dualism was easily rejected in principle, most researchers remained vulnerable to its lure in more implicit ways, in assumptions of what the brain did and where the cognitive operations took place (Bennett & Hacker, 2003). It might be true that thinking took time and left material traces, but this was a long way from explaining exactly what kind of cognitive operations took how much time and why.

Around the hundredth birthday of Donders's famous article, some researchers started getting more serious about deriving knowledge of the cognitive architecture from distributions of response times. Sternberg (1969a, 1969b) explored the use of search times in memory tasks to tease apart parallel versus serial processing. He asked his subjects to search their memory for items from a set they had learned by heart (or were trying to keep online in their head).

Some types of memory search gave flat response time curves, independent of the number of items in memory, whereas other types of search produced a steep increase in response time depending on the set size—the more items, the slower the response. The flat slopes suggested parallel processing, Sternberg concluded, and steep slopes indicated that the subject had to work one by one, serially considering each item in memory.

Treisman and colleagues (Treisman & Gelade, 1980; Treisman & Sato, 1990; Treisman & Souther, 1985) applied a similar logic to analyze response times in visual search tasks, and went one step further in the interpretation of the underlying cognitive operations with the feature-integration theory. Parallel search, seen in flat slopes, would occur for "singleton" targets, which differed from the distractors in only one visual dimension—like when you look for a red item among a set of green distractors. Serial search would be required whenever the target was defined on the basis of a combination of features—like when you look for a red square among blue squares and red circles. To combine visual features, you would need something called "attention" to glue the features together at one location at a time. Searching for a combination of features, then, meant that you would have to allocate attention to one location, let attention do its gluing there, decide whether the element at that location matches the target, and move on to the next location if it does not. The search time would literally depend on the number of times you have to shift attention to a new location.

The theory suffered badly from a load of incompatible data in dozens of new studies from other labs (see Wolfe, 2001, for a succinct review), but as a first shot it was not bad at all, and I have always admired its wonderful precision in translating response times to a fairly precise drawing of the underlying cognitive scheme. My own little experiment on visual search, in which I learned to wrestle with distributions, would have had no meaning if there were no feature-integration theory to shoot down.

Arguably the most forceful plea for response time analysis was put forward by R. Duncan Luce (1986), who had helped establish the threshold concept and was very familiar with the tenets of signal detection theory. He explained, for all who could follow, that it was feasible to exploit the shapes of response time distributions in an effort to deduce the covert operation of separable parameters relating to actions of the mind. For a long time, I thought this was the most esoteric of all things in psychology and statistics, something I would like to be able to understand if only I had the brain power for it. But then I encountered R. H. S. Carpenter's LATER model (Carpenter, 1981, 1999, 2004; Carpenter & Williams, 1995; Reddi & Carpenter, 2000), and I found myself actually coming to grips with it, or even liking it to the point that

I started working with it myself (Lauwereyns & Wisnewski, 2006). Here was a model that looked innocent enough, at least if you consider the schematic drawings, and it managed to work with just a handful of parameters, one which looked suspiciously like what I thought of as bias and another that just had to be sensitivity.

Carpenter wrote the most enjoyable introduction to the LATER model in his article for the *Journal of Consciousness Studies* (1999). The model starts from the observation that the eye movements of human observers are curiously slow if we consider the underlying anatomy for visual processing and eye movement control. Between the retina (when a stimulus excites the photoreceptor cells in the back of the eye) and the repositioning of the eyeball (when we move our eyes to bring the stimulus in central vision and examine it more closely), there should in principle be only a few synaptic steps involved, or a sequence of maybe five or six neuron-to-neuron transmissions. This should take up a few tens of milliseconds at the most. Instead, the response times with eye movements normally clock in at about two hundred milliseconds, and often even more. To explain this procrastination, Carpenter suggested, there must be a central decision-making mechanism that converts the available sensory evidence into an eye movement via a decision process with random variability. He even offered a philosophical perspective on the biological advantages of this random behavior, from escaping boredom and promoting creativity to outwitting our opponents and really willing freely.

Given the central role of random procrastination, the model is aptly named LATER. The acronym, however, stands for linear approach to threshold with ergodic rate, a rather ominous whole, in which I suspect the E was forced a bit for poetic reasons. Nobody really knows what "ergodic" means or whether it stems from "a monode with given energy" or "a unique path on the surface of constant energy" (Gallavotti, 1995). In practice, "ergodic" must be borrowed from the ergodic hypothesis in thermodynamics, which, brutally simplified, claims that, if you just measure long enough, you will find that a particle spends an equal amount of time in all possible states. In statistics, the ergodic hypothesis is taken to imply that sampling from one process over a very long period of time is equivalent to sampling from many instances of the same process at the same time. The process should be stable, no decay, no learning. With this caveat, then, the LATER model addresses decision making in a static context when the observer performs at full capacity.

The LATER model conceives of decision making as a process represented by a continuous, straight line, the "decision line," that rises to a threshold, or cutoff level—when the decision line crosses this threshold, the decision becomes effective, the motor execution is initiated, and the response time can

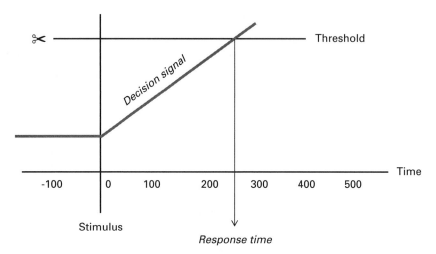

Figure 1.5
A linear model of response time in decision making. The horizontal axis represents time; the thick gray line gives the "decision signal" (a putative neural correlate of decision making). A decision is reached when the decision signal crosses a fixed threshold. In this example, the decision signal grows linearly from the time of stimulus onset and crosses the threshold in about 250 milliseconds.

be recorded (see figure 1.5). The model is based on just a few parameters: the starting point of the decision process (i.e., the distance between the intercept of the decision line and the threshold, assuming that the threshold is fixed); the average steepness, growth rate, or gradient of the decision process (i.e., the slope of the decision line—there is no shortage of synonyms); and the variance of the gradient.

The primary attraction of the LATER model is that it makes specific predictions about how changes to the parameters affect the shapes of response time distributions. With figures 1.6 and 1.7, I provide an unorthodox explanation that deviates substantially from the actual way in which the LATER model checks for changes to response time distributions. The true LATER model employs the reciprocal (or inverse) of response time—a little trick aimed at morphing the typically skewed response time distribution into a nicely symmetrical and normal one—and then draws the transformed distribution using a so-called "reciprobit plot," which pictures normal distributions as a straight line. The lines then swivel or move in parallel, depending on which parameter is changed. It works elegantly and makes for a straightforward statistical analysis, but to the untrained eye it can seem a bit confusing because a parallel change in the visual scheme of the model (moving the starting point up or down) translates into swiveling in the reciprobit plot, and vice versa. The

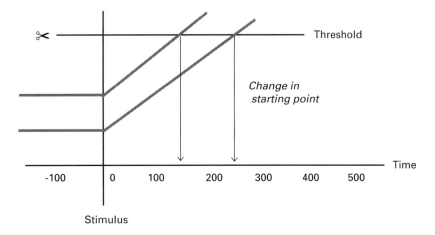

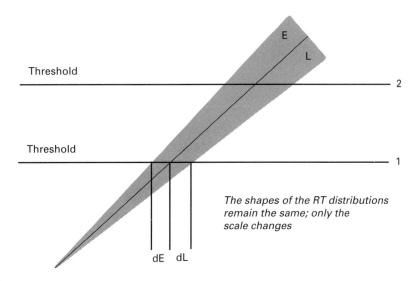

Figure 1.6
Effects of bias on response time. (a) The scheme is the same as in figure 1.5, including the decision line for the neutral case. Compared to the neutral case, the new decision line has a starting point closer to the threshold, allowing it to reach the criterion faster. Time is given in milliseconds. (b) Changing the starting point of the decision signal does not affect the ratio of the early (E) versus late (L) part of the response time (RT) distribution.

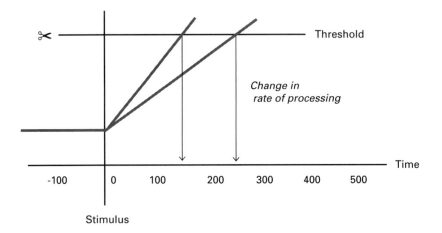

a

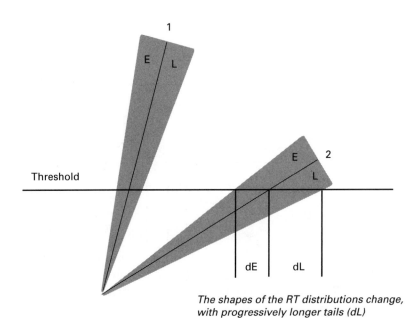

b

Figure 1.7
Effects of sensitivity on response time. (a) The scheme is the same as in figures 1.5 and 1.6a, including the decision line for the neutral case. Compared to the neutral case, the new decision line shows a steeper slope, allowing it to reach the threshold faster. Time is given in milliseconds. (b) Variation in the gradient (or steepness) of the decision signal influences the ratio of the early (E) versus late (L) part of the response time (RT) distribution: The tail of the distribution gets stretched out with shallow gradients.

problem is that the double transformation, using the inverse of response time and plotting in a funny way, warps the mind beyond our gut instincts (or implicit associations).

To satisfy my own intuitive inclinations, I tried to work out a rationale without the double transformation. Figure 1.6a shows what happens if we move the starting point closer to the threshold—a sure form of bias, or predisposition to reach a particular conclusion. It is easy to see that with the shorter distance to cover, response times will decrease considerably. Here, we go from 250 milliseconds down to 150. To get an idea of what this implies for the shapes of the response time distributions, figure 1.6b shows the decision line with its variance (the gray region; the thin line in the middle represents the mean). As we change only the distance between the starting point and the threshold, we might as well picture the situation with one distribution traveling up to threshold 1 or 2. Now we can consider how the early half of the response time distribution ("dE") relates to the late half ("dL"). The ratio of the two does not change. That is, the shape of the distribution stays the same; it only gets magnified if the decision line has to travel a greater distance.

Figure 1.7 applies the same logic for changes to the gradient of the decision line. Panel a shows what happens to the decision process if we have a steeper growth rate to the threshold, presumably due to more efficient information processing, or a clearer signal reception, that is, heightened sensitivity. With respect to the distance between the starting point and the threshold nothing has changed, but again we see a clear improvement in response time, from 250 milliseconds down to 150. Panel b works out the ramifications for the shapes of the response time distributions. Case 1, with a very steep growth rate, shows that the distribution approaches a symmetrical shape, with a ratio between the early half ("dE") and the late half ("dL") of not much less than one. Venturing into the absurd, we can even imagine a straight vertical decision line with a mean response time of zero, which would have a ratio between dE and dL of exactly one, or even more absurd, a line tilting to the left, with negative response times that imply a ratio of higher than one. Of course, in reality we can only tilt to the right, but the point should be clear: The ratio between dE and dL changes with the slope of the decision line. If we look at case 2, with a much shallower slope, we see that dL increases relative to dE, that is, the ratio of dE/dL dives well below one and the tail of the response time distribution gets stretched out.

However strange it sounds, or downright mystical, there must be some truth to Luce's (1986) dictum that we can read cognitive architecture out of response time distributions. With the LATER model, I found it was easy to relate mechanisms of bias and sensitivity to specific parameters that influence the shapes of response time distributions. This is not to say that everything is perfect with the LATER model. It cannot account for errors in decision making, and the idea that the decision process grows linearly must surely be a particularly vulnerable abstraction (see Smith & Ratcliff, 2004, and Bogacz et al., 2006, for comparisons of the strengths and weaknesses of different models). I would like to think of the LATER model as the simplest of all, and therefore the best place to start, even if it means tweaking the experimental paradigm so that errors are logically impossible (Lauwereyns & Wisnewski, 2006). But sooner or later, it may be necessary to extend the LATER model by adding parameters (e.g., Nakahara, Nakamura, & Hikosaka, 2006) or to develop nonlinear models that can account for error and exhibit a more neurophysiologically plausible growth rate—sometimes also called "drift rate" to emphasize that the rate does not necessarily grow (Ratcliff, Van Zandt, & McKoon, 1999).

Despite its limitations, however, the LATER model manages to provide an astonishing fit to response time distributions in some tightly controlled situations. In these cases, we can hope to apply the most powerful triangulation, measuring behavioral responses concurrently with neural activity on a trial-by-trial basis in one and the same experimental paradigm. Historically speaking, triangulation might have evolved as a method to measure the distance between shore and ship. Taking readings at two different angles on the shore, we should be able to work out where the lines will meet the ship in the distance. Applied to neuroscience, we can think of the experimental paradigm as the shore, and the behavioral and neural readings as our two angles that seek to meet the mind in the distance. No doubt most scientists will agree that this is the obvious best way to proceed. It is disappointing, however, how rarely it is applied in practice. All too often one of the two readings, usually the behavioral, is sketchier than it might have been.

The default approach seems to be to roughly compare one condition with another in terms of behavior as well as neural activity. Say we compare the ability of our observer to detect Jimmy Carter with glasses on versus off, and we measure the activity of a neuron in the observer's medial temporal lobe in the same two conditions. If we do what most researchers do, we will compute only four data points: the percentage of correct responses with glasses on versus off and the neuron's average Carter-to-noise ratio with glasses on versus off. Our observer makes fewer errors with glasses on and—lo and behold—the neuron fires more for Jimmy Carter than for anyone else... *We have a neural correlate of perception!* Or, the neuron fires less for Jimmy Carter than for anyone else... *We have a neural correlate of perception!* Whatever the neuron does, we will publish a paper in a very nice journal, but did we really compute a correlation?

We should be able to do better than that. We can record our observer's response times and use the trial-by-trial variability to check whether the gradient of his or her decision line is steeper with glasses on than with glasses off, as we might expect under conditions of heightened sensitivity. At the same time, we can check whether trials with shorter response times correlate with higher (or lower) firing rates for that wonderful neuron in medial temporal cortex.

The goal must be to establish the most detailed correlation of neural activity and behavioral responses, preferably in conditions that allow us to compute separable parameters in the responses. Then, and only then, can we hope to present a complete account of how neural circuits weigh the options. The best strategy will be to incorporate the LATER model or other tools of behavioral analysis in the design of experimental paradigms. As these tools become more sophisticated, researchers will be better equipped to examine, among other things, the neural signatures of bias and sensitivity in decision making.

Neural Signatures of Bias and Sensitivity

Combining the concepts of signal detection theory with the LATER model, there emerge a few solid markers that we can apply in the search for neural mechanisms of bias and sensitivity. At the moment, these markers are merely hypothetical, speculative, and conjectural, or subject to some of the most dreaded adjectives in science—they are the offspring of two very different ways of thinking about decision making, and so they may wither with the demise of either theoretical parent. Nevertheless, the markers are the proper kind of instrument for our search as they are wonderfully precise about what we should see in neural activity under the regime of bias versus sensitivity.

Figure 1.8 does the deductive work for the case of bias. In panel a, the data from a hypothetical neuron (or neural population) are drawn from a factorial design with 2×3 conditions: There are three possible types of stimulus (coded, similar, and other) and two possible treatments (biased or neutral). The different stimuli are needed to get an idea of what the neuron basically responds to—what kind of information does it normally "encode"?

The most thorough way to characterize a neuron's response properties would be by drawing a tuning curve in the way Vernon B. Mountcastle and colleagues originally conceived it (LaMotte & Mountcastle, 1975; Mountcastle, LaMotte, & Carli, 1972; Talbot et al., 1968), by systematically charting the changes in neural responses as a function of changes to a stimulus parameter. The stimulus that elicits the strongest level of neural activity, or the apex of the tuning curve, must be the prototypical stimulus, the one

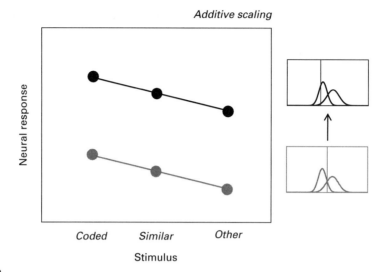

a

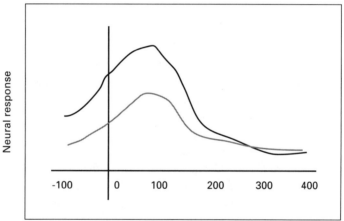

b

Figure 1.8
The neural signature of bias. (a) The horizontal axis of the main panel marks three different types of stimulus as a function of a neuron's basic tuning (or strength of response, listed from high to low): coded, similar, and other. The gray data represent a neutral case; the black data are driven by bias and show an additive increase as compared to the neutral case. The two inset figures to the right are borrowed from figure 1.3. (b) Average neural activity levels are shown over time (horizontal axis) relative to the onset of the stimulus (vertical line at time zero). The difference between the data driven by bias (black line) versus those from the neutral case (gray line) is already apparent before stimulus onset, reflecting anticipatory processing.

encoded, represented, or otherwise conveyed to the rest of the brain by the neuron in question—I call it the "coded" stimulus. (I prefer to avoid the term "preferred stimulus" because preferences sound too much like a matter of choice and the application of wishes, and that is not only too anthropomorphic for a neuron but also confusing with the real wishes in chapter 2.)

From early efforts in drawing tuning curves, it was immediately clear that, from the neuron's perspective, not all noise is equal. Tuning curves tend to show a peak that does not suddenly emerge from the flat but is supported by noticeable slopes, often in the shape of a bell. Panel a in figure 1.8 takes a shortcut, sampling just three positions on the tuning curve: the apex ("coded"), somewhere in the middle ("similar"), and the bottom ("other"), where noise is really just noise or maybe the perfect "antipreferred" stimulus, the exact opposite of what would get a neuron's juices to flow.

As an aside, we should note that, of course, the idea of a tuning curve should not be restricted to sensory stimulus coding. In fact, I prefer to think of tuning curves as part of the same family as receptive fields, mnemonic fields, and movement vectors, all the different charts and plots that characterize a neuron's firing rate with respect to any physical parameter in the experimental paradigm, be it spatial or nonspatial, present or past.

The gray data represent the neutral case. A shift of criterion, we noted, would be equivalent with a parallel movement for both the signal and the noise distribution in the framework of signal detection theory. Translated to the three positions on the tuning curve, this means we should see an additive increase when the observer is biased in favor of the "coded" stimulus that takes the apex of the neuron's tuning curve. The black data, driven by bias, seem to have undergone a parallel (linear) movement upwards from the neutral case, regardless of the actual stimulus, as if the tuning curve simply rides on top of an elevated baseline. This is, of course, also compatible with the LATER model, which further specified that the change in baseline, or starting point for the decision line, would be fully in place at the moment the first sensory evidence of the stimulus arrives. From this, we can distill an important second marker with respect to the temporal dynamics of bias effects. We should expect to find evidence of anticipatory processing, or a way in which the neuron effectively manages to change its "starting point" before the stimulus is presented. Figure 1.8b depicts a very visible way of elevating the baseline, with neural activity ramping up toward the expected arrival of a stimulus, more so when biased than when neutral.

Perhaps the most canonical way in which we can open the door for bias to influence decision making in a given experimental paradigm is to play with the probability of events. In doing so, we manipulate the prior probability, to use the terminology of Bayes's theorem. Carpenter and Williams (1995) had human observers make eye movements to peripheral visual stimuli during several blocks of trials. In each block, the likelihood that a stimulus would appear in any one trial was kept constant, from very likely (95%) to very unlikely (5%). The response time data matched nicely with the predictions of the LATER model, suggesting that with very likely stimuli, the decision line's journey toward the threshold was much shorter, and so the response much faster, than with very unlikely stimuli.

In the late 1990s researchers in several laboratories performed essentially the same experiment with monkeys while recording the activity levels of single neurons (Basso & Wurtz, 1997; Dorris & Munoz, 1998; Platt & Glimcher, 1999). In each case, the activity level of neurons was enhanced for stimuli or saccades whose prior probability was higher than that of other stimuli or saccades, even before the visual stimulus was presented (Dorris & Munoz, 1998) or before the monkey received an instruction about which of two possible stimuli was the actual target (Platt & Glimcher, 1999). The study by Platt and Glimcher deserves special mention as it showed data from three different experiments—not just the probability manipulation—and couched the entire data set in a then-unheard-of language, applying concepts from economics to the analysis of neural activity (read Glimcher, 2003, for the full introduction to the science of "neuroeconomics").

We will certainly have to return to the paper by Platt and Glimcher, but in the meantime, seeing as the Law of Eponymy is out of the window, we might as well highlight the massive contribution by Robert H. Wurtz, one of the authors of the cited 1997 paper on probability. In addition to mentoring a host of important researchers (including Michael E. Goldberg, Okihide Hikosaka, Douglas P. Munoz, William T. Newsome, Barry J. Richmond, and Marc A. Sommer, to name a nonrandom few), Wurtz compiled an impressive set of studies, showing time and again that neurophysiology—more specifically, the extracellular recording of action potentials from single neurons in awake and task-performing animals—can be applied with great effect to the study of elusive mechanisms operating somewhere in the big divide between sensory processing and motor control. The discoveries included "attention" (Goldberg & Wurtz, 1972), "memory" (Hikosaka & Wurtz, 1983), and "internal monitoring of movements" (Sommer & Wurtz, 2002).

For anyone who has witnessed or conducted this type of experiment, it is hard not to be amazed by the immediacy and precision with which it provides a window to the mental events that take place in the infamous black box, or the dark recesses beneath the skull. The first time I saw it, I was profoundly disoriented, unable to imagine how anyone could begin to invent a paradigm like that. In a brief and very readable story, specked with Nobel prizes, Charles Gross (1998) traced the historical origins of the technique back to Adolf Beck at the University of Kracow in the 1880s, who worked with rabbits and dogs, mapping visually evoked responses with field potentials in occipital cortex. It took another few geniuses, including E. D. Adrian and Stephen W. Kuffler, to move from field potentials to measuring things like "the receptive fields of single neurons in the cat's striate cortex" (Hubel & Wiesel, 1959; the paper is easily and freely accessed in digital form, courtesy of the *Journal of Physiology*). Gross's (1998) historical account ends at this point, but the science went on evolving.

By the 1960s, Edward V. Evarts (1966, 1968) was able to record from awake and task-performing monkeys. In September 1969, Wurtz published his first article in *Journal of Neurophysiology*, on the "visual receptive fields of striate cortex neurons in awake monkeys." It was the first of 66 in the same journal ("*The Journal of Wurtz*"), spanning four decades of total focus on the neurophysiological underpinnings of visual processing and eye movements in monkeys. Since the original paper in that special year of 1969 (the paper was published after, but submitted before, I was born), there registered no essential changes to the experimental paradigm: The monkey sat in front of a screen, looked for dots, and made eye movements, while Wurtz and his collaborators recorded the activity of single neurons.

Today, the technique remains arguably the most powerful method to study information processing in the brain, providing a temporal and spatial resolution far beyond what can be reached with other methods while subjects are making decisions. In principle, the technique allows researchers to compute trial-by-trial correlations between behavioral response times and neural activity, measured on a continuous time scale (down to milliseconds, enough to pick up each individual action potential) and at the level of single neurons (down to micrometers). Much of what I have learned about neural mechanisms of decision making is based on the firing rates of individual neurons, and so Wurtz-like papers (more commonly called "single-unit studies") will feature heavily among my references. This is not to say all is well with the technique.

A cautious ethical note must be attached. The invasive nature of the technique drives researchers to work with animals other than humans—a move that is not appreciated by everyone in the same way. The present monograph is hardly the place to elaborate on the issue, but the minimal stance, implied also in the U.S. Animal Welfare Act, should be to look for alternatives wherever possible. Whether other techniques are viable replacements depends on the topic under investigation and on the level of precision required. In some cases, we can take brain scans to measure the cerebral blood flow in humans as they perform tasks, via functional magnetic resonance imaging (fMRI) or positron emission tomography (PET). Especially, fMRI has come to the fore quite vigorously in the past ten years or so. There will be a good portion of fMRI studies among my references as well. With fMRI, we trade temporal resolution (in seconds) as well spatial resolution (in millimeters) for a major improvement in external validity—working with the right species, drawing no blood, and applying decision-making tasks that go from anything a fruit fly can do to things that only the smartest of us can do. Though the relation between cerebral blood flow and neural activity is yet to be determined precisely, there can be no doubt that the so-called blood-oxygen-level-dependent (BOLD) signal in fMRI does in fact provide a reliable parametric estimate of the extent of neural processing in a given brain structure (see Logothetis, 2008, for the state of the art).

In other situations, we might wish to measure "brain waves," or global electrical activity, from the scalp, via electroencephalography or its newer cousin, magnetoencephalography, with good temporal resolution (down to milliseconds) but poor spatial resolution (at the level of entire lobes at best). We can also learn a great deal from how the brain responds to drugs or more damaging assaults, either induced experimentally in an animal or occurring naturally as when one of us suffers a stroke or gets injured in an accident. Some kind of convergent approach seems the obvious best solution, working from multiple angles and with different methods simultaneously. Neuroscience certainly benefits from its wide variety of research tools and paradigms, and I will draw on any of them in my attempt to provide a coherent account of how neural circuits underscore decision making.

Coming back to the Wurtz-like studies, one recent development that certainly has my sympathy is a gradual shift toward a different species—monkeys still dominate the decision-making scene, but rats are gaining fast (e.g., Houweling & Brecht, 2008; Kepecs et al., 2008; Pan et al., 2005; Roesch, Calu, & Schoenbaum, 2007). Switching to rats creates a magnificent opportunity for a more integrated systems-neuroscience approach, including pharmacological, anatomical, genetic, and intracellular electrophysiological techniques that are too costly with monkeys, in both ethical and financial terms. Relying on single-unit studies with rats versus fMRI studies with humans, it should be possible to significantly reduce the future need for monkey research. At present, however, we should fully acknowledge the crucial role of monkey single-unit studies in the accumulation of our database on the neural mechanisms of decision making.

Thus reinvigorated, we pick up the paper by Basso and Wurtz (1997) again, and appreciate its demonstration of how "the role of the prior" is translated into systematic variation of anticipatory processing in individual neurons. The monkey was required to make an eye movement to a visual target that could appear at one out of a predetermined set of possible locations, indicated by placeholders. The set size changed on a trial-by-trial basis, from one to eight possible locations. Basso and Wurtz recorded from buildup neurons in the superior colliculus (a type of neuron first identified by Munoz & Wurtz, 1995). Was it a strategic choice to record from buildup neurons? In hindsight, it definitely seemed the perfect pick. If we would like to find evidence for bias in anticipatory processing of a form like that presented in figure 1.8, then it makes total sense to focus on neurons that naturally tend to ramp up their activity levels in preparation for task events. And, sure enough, the baseline activity of a typical buildup neuron, in response to a placeholder in its receptive field, increased as target uncertainty decreased, well before the actual target appearance.

More recently, similar manipulations of target uncertainty have yielded the predicted type of differences in the baseline activity of other neural structures not specifically associated with buildup activity. A single-unit study in monkeys showed the effect in lateral intraparietal cortex during a motion-discrimination task (Churchland, Kiani, & Shadlen, 2008), whereas an fMRI study showed increased activity in extrastriate and anterior temporal lobe regions when human observers needed to compare the orientation of Gabor patches against one alternative rather than two (Summerfield & Koechlin, 2008). The effects of probability, then, do seem to accord with the neural signature of bias as pictured in figure 1.8. The search is on for other determinants of bias.

As a corollary of the studies on probability, there appears to be a conspicuously systematic, inverse relation between the number of alternatives and the ease of decision making. In signal detection theory and the LATER model, however, the actual number of alternatives seems to be abstracted away, as decision making is translated into a contest between signal and noise—to be or not to be, in the parlance of the Prince of Denmark. In defense of the reductive attitude, we could point to the etymology of the very word "decision," from the Latin *decidere*, or *de* + *caedere*, "to cut off." It may not be entirely clear what the ancient Romans were in the business of cutting off, but I would prefer to take the least bloody interpretation, as in putting an end to nothing more material than a thinking process, or the internal agonizing over different options. Perhaps the threshold theory really is thousands of years old. In any case, the general implication seems to be that decision making is all about reaching something final or definitive, a conclusion, a solution, an outcome, a statement, a proposition, a value on the color map, a number from one to a hundred, a judgment of character, a sentence with three subordinate clauses, some *thing*, of which there usually is only one, even if it is a highly convoluted one, or something tricky, operating on a metalevel, like the decision that there will be no decision, as when a legal court rules that it has no jurisdiction over the matter at hand.

The rigorous one-track mind, holding onto the necessary singular of *the* decision, can capture the weighing of multiple alternatives as a parallel competition among different "decision units"—as if we have not one but several LATER functions running at the same time, one for each alternative, such that what constitutes a "signal" for unit A is actually part of the "noise" for unit B. As soon as one of the LATER functions reaches its threshold, it would take home the big prize, and be proclaimed the winner—the decision outcome. Or maybe we employ another algorithm to identify the solution. Or maybe the different decision units interact or influence each other. In any case, the notion of different decision units, working in parallel or interactively, helps us bend the two-choice logic of signal detection theory and the LATER model to suit the needs of any decision-making situation. In chapter 6, "Less Is More," I will return to the multiplicity of decision processes at any one point in time, no matter how dormant or awake they are.

However, now it is about time we take a look at figure 1.9 and learn to recognize the neural signature of sensitivity. The presentation format is the same as that of figure 1.8, with a hypothetical neuron in a factorial design. In panel a, we have the three types of stimulus (coded, similar, and other), and two treatments—this time, increased sensitivity (black data) versus neutral (gray data). Increased sensitivity is achieved by reducing the overlap, or widening the distance, between the signal and noise distributions according to the proposals of signal detection theory. Applied to the tuning curve, this produces an enlarged ratio of the response to the "coded" stimulus relative to any "other" stimulus. Put differently, in absolute terms the effect of sensitivity on neural firing should be larger for the "coded" stimulus than for any "other" stimulus. This corresponds to a multiplicative scaling effect, as if the tuning curve is multiplied by a constant sensitivity factor greater than one—effects of this kind are sometimes tagged as "gain changes," although the term "gain" remains ambiguous with respect to the additive versus multiplicative nature of the effect, blurring the difference between bias and sensitivity.

Distinguishing between these two mechanisms, as do the LATER model and signal detection theory, is a sine qua non if we wish to unravel how the brain provides us with the computational power to make decisions. Distinguishing between additive and multiplicative effects on tuning curves should be very useful indeed and might surely be practiced more often. Of course, the additions and multiplications will rarely work out perfectly in the quirky

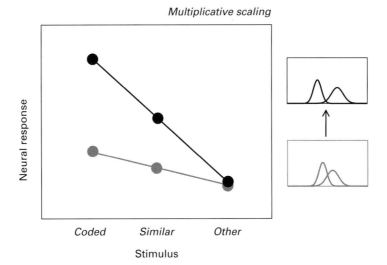

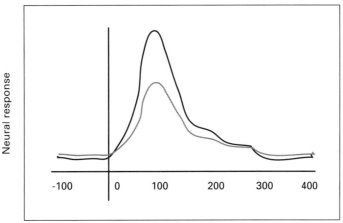

Figure 1.9
The neural signature of sensitivity. (a) Here, the black data, driven by increased sensitivity, show a multiplicative effect as compared to the neutral case. The two inset figures to the right are borrowed from figure 1.4. (b) The temporal dynamics of the neural response to a "coded" stimulus. The difference between the data driven by increased sensitivity (black line) versus those from the neutral case (gray line) emerges only after stimulus onset, reflecting synergistic processing.

reality of empirical data. For instance, bias might add a smaller absolute amount for the "coded" stimulus than for "other" stimuli when the neural firing rate hits its maximum capacity. The addition is then obscured by a "ceiling effect," leading to a signal-to-noise ratio that would even deteriorate under the bias regime as compared to the neutral case. Conversely, sensitivity might outdo multiplication and show exponential growth. Nevertheless, such deviations and complications can be formulated precisely in computational terms. The main point is that the logic of bias versus sensitivity does translate into differential movements on tuning curves. The peculiar fluxions are there for us to check up on in the data and to use strategically as markers and diagnostics for the involvement of this or that underlying neural mechanism.

In figure 1.9a, then, we note that the black data, driven by increased sensitivity, swivel upwards from the neutral case, with a degree of change that depends on the actual stimulus being presented. Here, there must be some kind of ad hoc interaction between the signal processing and the mechanism of sensitivity. This makes perfect sense, of course—the changed signal-to-noise ratio can only become visible when there is, in fact, a signal to be processed in the first place. In the LATER model, we can easily appreciate what the interaction does for the temporal dynamics of sensitivity effects. The steeper gradient can only take effect once there is some sensory information to work with, that is, from the moment of stimulus onset. The incoming sensory information and the increased sensitivity work together, simultaneously, interactively, synergistically—taking "synergy" in its early sense, derived from the Greek *sunergos*, "working together," instead of the "rather blowsy word" it is "these days, with its implications of corporate merger for profit-enhancing capacity" (dixit the word-cleaning poet, Michael Palmer, 2008, p. 28). Figure 1.9b shows how the neural response to a "coded" stimulus reaches a much higher amplitude with increased sensitivity (black line) as compared to the neutral case (gray line). Yet, the difference in the neural response emerges only after stimulus onset.

In practice, we can easily examine the neural correlates of increased sensitivity by physically modifying the degree of similarity between signal and noise. The most thorough investigation of this kind was, and is, being conducted using a perceptual discrimination task with different levels of motion coherence, in which the subject has to report the dominant direction among a set of moving dots. The task is easy enough when all dots move in the same direction (100% coherence) and obviously is impossible when the dots move in random directions (0% coherence). Between these two poles, performance improves steadily with higher coherence levels. In terms of response times, the improvement should be attributed to changes to the gradient of the decision line in the LATER model, as confirmed in a study by Reddi, Assress, and Carpenter (2003). However, the coherence levels also affect neural activity in exactly the way we would expect, with neurons reaching higher firing rates for easily discriminated stimuli that match the "coded" direction. (In addition to the already cited single-unit work by Churchland, Kiani, & Shadlen, 2008, in which probability and motion coherence played in concert, the landmark studies employing motion coherence were performed by Britten et al., 1992; Newsome, Britten, & Movshon, 1989; Roitman & Shadlen, 2002; and Shadlen et al., 1996; for an fMRI version, see Heekeren et al., 2006.) The neural signature of increased sensitivity bears out perfectly in these data, both the multiplicative scaling and the synergistic processing.

Arguably the most thorough analysis of increased sensitivity in neural firing rates as well as response times was performed by Ratcliff and colleagues (2007) on the basis of data from superior colliculus neurons while the monkey performed a brightness-discrimination task, in which some levels of brightness were easy to discriminate (98% white pixels, very "bright," or 2% white pixels, very "dark"), others hard (45% or 55% white pixels). Response times were fast and neural firing rates high for easy "coded" stimuli, and again the effects in neural processing emerged only after stimulus onset. But more than this, the study reached an unprecedented level of detail in modeling the trial-by-trial variation of both response times and neural activity—a great achievement, exactly the type of triangulation that forms my ideal of neuroscience.

Armed with the analytic tools to distinguish bias versus sensitivity, familiar with the Bayesian way of thinking about decision making, no longer afraid of signal detection theory and the LATER model, always on the lookout for anticipatory versus synergistic processing, and eager to compare additive versus multiplicative scaling, we are now ready to investigate how neural circuits really weigh the options, in what kind of conditions, under what sort of circumstances, and to what degree of inevitability. Having duly sniffed at the formulas, we can finally take a look at how they function.

2 Wish Come True

Yearning for the Emperor Tenji
While, waiting for you,
  My heart is filled with longing,
The autumn wind blows—
As if it were you—
Swaying the bamboo blinds of my door.

Princess Nukada (or Nukata) wrote this little poem close to a millennium and a half ago (Manyōshū, 2005, pp. 11–12). Most Japanese readers today cannot read the original—it took a team of half a dozen scholars to come up with a decent paraphrase in contemporary Japanese, and then another team of three translators to turn the text, by now properly stripped bare of its poetry, back into something that sounds lyrical enough in English. It should have been an impossible task, but oddly, for me, the end result works beautifully.

The words do capture my imagination and bring to life some kind of deeply emotional scene, however remote from historical actuality. The knowledge that the poem is a precious relic of a different age likely contributed to my receptiveness. The figure of Princess Nukata is properly enigmatic, regarded as the greatest poetess of her time. She was yearning for one Emperor in the quoted poem but later became the wife of another (the younger brother of the first). I first encountered her name, with a letter "t" instead of "d," in a fascinating poem by Sadakazu Fujii (published in Fitzsimmons & Gozo, 1993). Fujii mixed the past with the present at a conference on Japanese classics and ended up desperately pining for his own Nukata, a new incarnation of *la belle dame sans merci*, in the words of John Keats.

The intertextual play surely must have magnified my sense of occasion, my readiness to fully read into the desire, the longing, and the family of related feelings that form the object of the tiny ancient poem. A dyed-in-the-wool post-postmodern literary theorist might scorn me for my surprise at being swept so easily from my feet by the words of Princess Nukata (others might simply call me sentimental). With Jacques Lacan's dictum that desire only fully comes into existence in language, I might have expected that the pangs of Nukata's yearning would be felt all the more under extra layers of words—or did I misread the Frenchman? (Try Lacan, 2004, for a sample of the dense, but wonderfully cadenced sentences that proclaim their ideas in a magnificent kind of bizarre complexity, more poetry than science.)

Does desire really belong to language? I can easily see that symbols and abstract representations are tools that help us maintain virtual copies of things in mind, especially useful when the real objects temporarily disappear from view but might very well return sooner or later. The negative corollary comes into play when these virtual copies move about freely in mental space while their actual referents are in fact beyond returning, as when Princess Nukata wishes for the impossible upon hearing of the (first) Emperor's death. The mere fact that we are endowed with a faculty like that of language, making us aware of what we miss, puts us at risk for desperate desire, unrequited love, and impossible dreams of bringing our loved ones back from the dead—exactly what Orpheus, the mythical first poet, tried to do for his Eurydice. Without his poetry, Orpheus might never have missed her. But then he went on to do what he was not supposed to do; he failed to inhibit an eye movement, looked back, and saw his Eurydice retreat forever in the darkness of the underworld. His poetry did not help him there either.

Another Frenchman of the psychoanalytic bent, Maurice Blanchot (1989), pointed to the mythical moment of Orpheus's disastrous saccade as the quintessential image of what the space of literature is all about, from Eros to Thanatos, in a way that sounded not unlike the intuitions of the true master of psychodynamics in *Beyond the Pleasure Principle* (Freud, 1990). But having reduced all of the human condition to a single, fixed sequence of two highly abstract instincts, and having installed language as the categorical border between the psychological and the biological, have we not gone off the deep end, lost control of our monomania, and given in to the very subjective nature of desire, claiming it is so personal it can only be ours? We might as well say that without language there is no subjectivity, no desire, nothing—or, to quote the one biblical verse I know by heart, "In the beginning was the Word" (John 1:1).

Perhaps the bible and Jacques Lacan provide us with an important lesson about language, about the awe that it inspires and the magic that it achieves, as a device that organizes our experience and works in ways that are not bounded by physical constraints. Yet, to move on from this observation to claiming that language is a necessary condition for things like consciousness, desire, and subjectivity looks to me like an illicit shortcut to preventing an independent examination of the very phenomena that are seen to be contingent on language. If things do not exist outside language, they can only be investigated through language—through practicing the talking cure, chatting interminably or writing poem after poem, much like humans did for centuries with great emotional effect but without getting any closer to an actual understanding of the experiences that are shaped by the words.

I think I smell a tautology here, a petitio principii, the hard-core Lacanian intellectual begging the question of what desire really is, fencing off the concept in absolutist terms that rule out the possibility of evidence to the contrary. With Popper (2002a), I would prefer something falsifiable, open to investigation. Without throwing Lacan's ideas out the window entirely, we might see some kind of vicious (or virtuous, if you will) circle at work, desire finding an expression in language, which reinforces the desire, only to turn back to words in a more complex and idiosyncratic format, and so on. The starting point, though, need not be language.

In fact, we might hypothesize that there exist simpler forms of longing and yearning, needing and wanting—prospective mechanisms that exist before language but are structurally directly related to the more complex, linguistically mediated versions. This hypothesis can be checked anatomically, neurophysiologically, and psychologically. It invites researchers to devise experimental paradigms that precisely define prospective mechanisms, from the prelinguistic to the most complex, in measurable terms. The experiments will generate data for comparison, allowing us to look more closely and open-mindedly at the differences and similarities among different mechanisms. With any luck, we might actually learn something.

Heeding, once more, the words of Kenkō, our favorite Zen Buddhist priest, we do well to forget about the intertextuality for a minute and look at Princess Nukata's poem for how "the lovely purity of the form creates a powerful impression" (Kenkō, 2001, p. 13; writing seven hundred years forward in time from Princess Nukata, he may have had specifically her poems in mind). The poem's emotional color is explicit and unambiguous, "yearning" being the first word in the title, followed by five lines that do not question the internal state but illustrate its intensity by zooming in on a wish mistakenly thought to have come true: "The autumn wind blows—/As if it were you—/Swaying the bamboo blinds of my door" (Manyōshū, 2005, p. 12).

The sound of the wind, and how it shakes the bamboo blinds, leads Princess Nukata to make a false alarm, to speak with signal detection theory—an erroneous interpretation of sensory information that she might not have fallen for if she had recently quarreled with Emperor Tenji upon catching him with his kimono open in the arms of another Lady of the Court. In the case presented by the poem, the yearning produced a bias in decision making, Princess Nukata all too eagerly accepting noise to be her desired signal. Fans of Lacan might argue that the true concept of desire only kicked in once Princess Nukata became aware of her false alarm, recognized its absurdity, felt a pang of regret, and had to write a poem about it. However, as a place to start investigating the entire complexity of the scene, I think we had better look at the basic process of wishing things to come true and how this affects decision making. How does the wishing begin? Where does it start?

What Is inside the Skin?

I borrowed the title for this section from Burrhus Frederic Skinner's (1974) *About Behaviorism*. In many ways, we can think of Skinner as the inverse of Lacan, the exact opposite, the total antipodean—the two were each other's contemporaries, more or less obliquely acknowledged each other's existence, but basically moved in separate universes, further polarizing the divide that existed already between thinking in French versus English, conceptually versus empirically. Lacan led the critical second movement in psychoanalysis, securing its survival after Freud, bringing it closer to philosophy and linguistics and further away from biology. Skinner provided the most complete definition of behaviorism as a fundamentally empirical science, taking extreme care to connect concepts with physical phenomena. Cartesian dualism had never dominated the field as thoroughly, if implicitly, as when Lacan and Skinner ruled the primetime media in academia—Lacan in charge of the errant processes of *res cogitans* and Skinner reigning supreme over the architecture of reflexes in *res extensa*.

Or am I wrong? Did I buy into caricature representations of both Lacan and Skinner in textbook summaries? Lacan's position on language and desire, for instance, sounded quite a bit more attractive with just a subtle change to the proposal, moving from begging the question to reinforcement in a vicious circle. Perhaps that was what Lacan had envisaged all along. What about Skinner? Did I fail to understand that Skinner really did want to talk about the mental? *About Behaviorism* took me by surprise when I actually read it for the first time, not too long ago.

Skinner's writings are positively entertaining, silky, and effective, as with the puns on his own name in the titles for his second and thirteenth chapters ("The World within the Skin" and "What Is inside the Skin?" the former sounding like an answer to the latter). I think they also give a cheeky little playful nod to his caricature self as a skinner, an extreme reductionist, who is only interested in the animal's outside for the most materialistic purposes of producing leather, making shoes, and earning a pittance. His *Introduction* lists twenty common misconceptions on behaviorism, such as "It assigns no role to a self or a sense of self" (Skinner, 1974, p. 4) or "It regards abstract ideas such as morality and justice as fictions" (p. 5). I had to plead guilty, and started reading properly.

Instead of blind reductionism, I found a radical commitment to conceptual criticism, favoring parsimony and clarity—an attitude that I could not judge to be objectionable. A minimalist empiricist who thinks lucidly there might be a self, concerned with the reality of abstract ideas such as morality and justice, possibly vulnerable to a vicious circle between language and desire...Could it be that Skinner wanted to be able to talk with Lacan after all? If Skinner did not, I do. To cover the distance between the two, I would move from Skinner to Lacan, not the other way around. I would begin my investigations with the basic processes of wishing things to come true in their simplest form of something prospective in relation to a target. With Skinner (1974, p. 107):

To look for something is to behave in ways which have been reinforced when something has turned up. We say that a hungry animal moves about looking for food. The fact that it is active, and even the fact that it is active in particular ways, may be part of its genetic endowment, explained in turn by the survival value of the behavior, but the way in which an organism looks for food in a familiar environment is clearly dependent upon its past successes. We tell a child to find his shoe, and the child starts to look in places where shoes have been found.

At first sight, this description of the search processes goes the full distance to avoid any consideration of the biological mechanisms and cognitive processes that might be at work inside the skin. The tone seems completely in accordance with the common (mis)conceptions about behaviorism. Apart from the brief mention of a "genetic endowment," there is nothing to suggest that the search function is intrinsically connected to the activity of neurons or the maintenance of internal representations. There is merely behavior that tends to reoccur after success, whether it be performed by an early model of the hunter–gatherer (pictured in figure 2.1) or by Elmer and Elsie, the tortoise-like autonomous robots that search for battery power (specimens of the species "Machina Speculatrix," first constructed by W. Grey Walter in the 1940s; see Steels, 2003, for an inspired review on behavior-based robotics and the dynamic road to intelligence).

In the language of Skinner (1974), the forward locomotion of the toddler may very well functionally be equivalent to that of the artificial tortoise. Of course, this is precisely the point of the minimalist approach, trying to work from the bottom up, from the simple to the complex, as far as possible, until

Figure 2.1
"We tell a child to find his shoe, and the child starts to look in places where shoes have been found," said Skinner (1974, p. 107). Here, the search behavior in question, performed by an early model of the hunter–gatherer, appears to be interrupted by a parental version of the observer's paradox.

we encounter an obstacle, a wall, or a dead-end street. It is only when the toddler does something altogether different and unexpected that we may have to add another circuit inside the tortoise until it does not work anymore and the toddler wins. When artifacts and computational models fail to perform the functions observed in biological organisms, we might learn something important about the boundary conditions for the strange phenomena of life.

Yet, however low on fat the quote from Skinner (1974) may be, there do sneak in two very important proposals that provide a powerful framework for investigation. Genetic endowment would be related to the survival value of behavior, and, in a familiar environment, current behavior would depend on previous successes. Nature and nurture both influence behavior on the basis of past behavior but work on different time scales. Nature works the gene pool across the generations. Nurture operates within an individual's lifetime. By formulating both as historical processes with respect to behavior, Skinner in effect urged us to see the parallels and think in Darwinian terms about shaping behavior through selection regardless of the substrate.

Taken to be an algorithm operating on behavior, "Darwin's Dangerous Idea" (the way Dennett, 1996, called it) certainly has a lot of potential as a theoretical framework to study the shaping of behavioral processes. Skinner (1974) did not harp on the thought as explicitly as Edelman (1987, 1992) would for neurons and cognitive processing a few decades later (or Dennett, 1991, or Calvin, 1996). Nor was Skinner the first to do the harping. That feat must probably be attributed to the inevitable William James, dixit Calvin in *The Cerebral Code*. However, Skinner (1974) unmistakably did connect the idea of behavioral success with selection of the successful behavior, inserting Darwinian processes in the day-to-day, or even trial-by-trial, variability of a subject's doings.

Implied in the behavioral dynamics, then, is some historical trace, a pruning or weighting of mechanisms as a function of behavioral success, which would effectively implement a bias in the search process. The mechanisms pruned or weighted must include some crude form of representation, of behavioral options in a given context, even if we would view the representations primarily and perfectly robotically as "organizers of activity rather than abstract models of some aspect of reality" (Steels, 2003, p. 2390). Stated in this way, it is only a very small step indeed to move from the history of behavioral success to investigating how the neural circuits of decision making undergo changes that correlate with the behavioral dynamics.

What remains to be defined is "behavioral success." In keeping with the Darwinian perspective, we might broadly relate behavioral success to an increase of inclusive fitness—a concept introduced by William Hamilton (1975), and popularized by Richard Dawkins in *The Selfish Gene* (2006), as a more general evolutionary model than kin or group selection. Fitness and survival would go beyond the self or the individual and include the next of kin or other social partners, explicitly grouped or not. Some outcomes are good for an individual's fitness or for the fitness of others around the individual, as measured most basically in terms of the ability to propagate genes, with implications fanning out in all directions, from the state of an organism's internal milieu and the maintenance of energy resources crucial for survival to the ability and opportunity to reproduce once you are finally sexually selected by your loved one. In humans we may expect the entire fitness picture to be particularly complex as it involves many layers of abstraction and intervening memes (a term coined by Dawkins in the very same *The Selfish Gene*), from mental health to social reputation, and from political power to cultural norms.

Outcomes that promote inclusive fitness would "reinforce" behavior, strengthen it, or make it more likely to reoccur on the next occasion when the individual is confronted with a similar set of circumstances. Edward Thorndike

(1898) was the first to describe this process of behavioral adaptation, commonly referred to as "instrumental learning" or "operant conditioning." He introduced the Law of Effect to capture the feedback mechanism, including intermediate variables such as pleasure or satisfaction that sounded too mentalist to Skinner's ears. Doing away with the internal states—not to deny their existence, but to break down the process to its bare essence—Skinner (1974) focused on the physical parameters of input and output and made a crucial distinction between positive and negative reinforcement: "A positive reinforcer strengthens any behavior that produces it" whereas "a negative reinforcer strengthens any behavior that reduces or terminates it" (p. 46).

I have always found this a bit confusing, and preferred to think of rewards and punishers, things we like versus things we do not. Of course, Skinner's point was to remove the intervening variable of "liking," and to do away with the assumption of agency in the act of punishment. Another, more "natural" way of talking might be to suggest that likable things are those we would approach, and aversive things those we would escape from, but even this little bit of vocabulary implies implications that do not hold up, as when we employ approach behavior to create an escape—offense sometimes being the best defense, or swatting mosquitoes the most effective way to avoid a bite.

Still, now that Skinner (1974) has done the hard, minimalist work, making explicit the myriad assumptions and implications that we should be aware of, we can try to come back to English and save Skinnerese for technical reports. Thus, wishful thinking would be about approaching rewards—and a reward would simply be the following as per pocket Oxford (Thompson, 1996, p. 779): "n. 1 a return or recompense for service or merit." What does in fact count as recompense? In the most basic sense, it should relate to an organism's inclusive fitness, determined in part by its well-being. Even if Skinner did all he could to avoid internal states, there must be some way in which the outcomes feed back to the mechanisms that make up an individual's behavior in a way that reflects on the issue of well-being in the first place. Whichever way we look at it, the question of life involves some kind of economy with respect to the body. The same object (a glass of water) might or might not positively reinforce depending on the state of the organism (thirsty or sated). How can the state of an organism regulate the reinforcing values of different objects? At some point, the body must have a say about what is useful for its maintenance.

Body states and the regulation of life, however, would be the perfect breeding ground for a few extra loops between input and output. With a little recursion and representation, the computations of the values of objects and events would inevitably start looking suspiciously like mental states. Before we know it, we will find ourselves feeling feelings. If anyone deserves their name to be attached with this scenario, it must be Antonio Damasio, who developed the somatic marker hypothesis (Damasio, Tranel, & Damasio, 1991) and the Iowa Gambling Task (Bechara et al., 1997) and wrote several inspiring books about the brain, consciousness, and emotion (Damasio, 1999, 2003, 2005). Here is a relevant quote from *Looking for Spinoza* (Damasio, 2003, p. 111):

> Feelings probably became possible because there were brain maps available to represent body states. Those maps became possible because the brain machinery of body regulation required them in order to make its regulatory adjustments, namely those regulatory adjustments that occur during the unfolding of an emotional reaction. That means feelings depend not just on the presence of a body and brain capable of body representations, they also depend on the prior existence of the brain machinery of life regulation, including the part of the life-regulating mechanism that causes reactions such as emotions and appetites.

Damasio (1999) makes a somewhat tricky distinction between emotions and feelings, the latter being more complex, conscious states, sounding dangerously close to a categorical divide, not unlike some of the proposals by Lacan (2004) or even the very Descartes (1993) thought to be in error. However, if we work with numbers of reverberations among brain maps, lengths of loops, and sizes of circuits, we can look for measurable quantitative differences among the various physiological properties that underscore internal states, even if they add up to become the whole of a feeling that is qualitatively different from an emotion.

Of course, the idea that reverberation and looping is the way forward from an emotion to a feeling, or from an unconscious internal state to a fully conscious and linguistic one, is stubborn enough to return in several guises, be it an eternal golden braid (Hofstadter, 1999) or the austere acronym "FLN" (faculty of language in the narrow sense, or simply recursion; Hauser, Chomsky, & Fitch, 2002). Whichever form the hypothesis takes, it boils down to a fundamentally *biological* explanation of the mind, rejecting the traditional, creationist view of the body as a vehicle purposefully designed to carry the soul, and offering a neutral, historical view instead, with the mind emerging as a function of the body, originally fully occupied with basic questions of survival and gradually transforming, if not transcending, into more sophisticated forms of thought.

Along this historical process, the outward shape of the rewards and recompenses that occupy the mind would change as well, no longer merely or exclusively the basic objects needed for survival or inclusive fitness, from food and shelter to a busy mating season, but also the more elusive and abstract objects, from money and fame to the really arcane and Hegelian ones of gaining knowledge for the sake of knowledge and appreciating the wonderful complexities of the world in an aesthetic, religious, or philosophical frame of mind. However, even if the rewards and recompenses are categorically different for fruit flies and rats versus philosophers, the underlying neural circuitry of reward-oriented behavior would share important commonalities in the way neurons interact, synapses are formed and reformed, and ions are exchanged. The biological hypothesis invites researchers to actually roll up their sleeves and investigate how rewards influence neural circuits.

A Place in the Skull for All We Know

Applying the title of a movie with Montgomery Clift and Liz Taylor to the concept of "a skull place," borrowed from the Dutch poet H. H. ter Balkt (2000), there can be no doubt that for all we know (sung, preferably, by Nina Simone) there must be a place in the skull for everything we do have knowledge about (see Boden, 2008, for hands down the most impressive historical account of how this body of knowledge came into being). Though it may be easy enough to establish that the mind somehow goes together with the brain, rather than the heart or the kidney, it is quite a different task to disambiguate the *somehow*. Karl Lashley (1931, p. 245) noted the following in his classical paper "Mass Action in Cerebral Function":

No one to-day can seriously believe that the different parts of the cerebral cortex all have the same functions or can entertain for a moment the proposition of Hermann that because the mind is a unit the brain must also act as a unit.

But even so, the deviation from an unreferenced Hermann did not imply giving in to anything as simple as cerebral localization with a one-to-one mapping of cognitive function onto neural structure. One page further (or several minutes later, during his delivery of the lecture before the Harvey Society in New York on November 20, 1930), Lashley remarked:

[T]he classical concept of cerebral localization is of limited value, because of its static character and its failure to provide any answer to the question of how the specialized parts of the cortex interact to produce the integration evident in thought and behavior. The problem here is one of the dynamic relations of the diverse parts of the cortex, whether they be cells or cortical fields.

Lashley as the iconic antilocalizationist sounds perfectly in sync with contemporary ideas about systems neuroscience, but his 1931 paper went on to formulate the mass action in cerebral function in a way that seems no less opaque than the unitary mode of the maligned Hermann. Lashley no doubt over-emphasized the size of cortical fields as a determinant of how the brain–mind works. Neuroscientists today prefer to think of dynamic processes gated through specific neural circuits, mapping cognitive functions onto sequences of activity across weighted networks of neurons. The place in the skull for anything we know would be a symphonic composition performed by groups of neurons—a spatiotemporal pattern moving in milliseconds inside the brain.

Depending on which neuron we listen to, we will hear a different fragment of the composition. Figure 2.2 presents a very sketchy sketch of the human

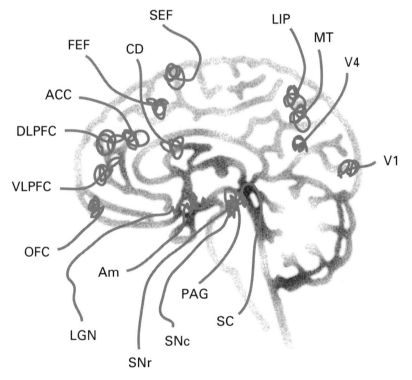

Figure 2.2
Rough sketch of the human brain. Indicated are some of the major structures for decision making, as well as more basic structures for visual processing. The sketch shows a midsagittal slice (cut straight through the middle of the head), front to the left and back to the right. The more or less indicated structures are (in clockwise order, starting left): VLPFC, ventrolateral prefrontal cortex; DLPFC, dorsolateral prefrontal cortex; ACC, anterior cingulate cortex; FEF, frontal eye field; CD, caudate nucleus; SEF, supplementary eye field; LIP, lateral intraparietal area; MT, middle temporal area; V4, fourth visual area; V1, primary visual cortex; SC, superior colliculus; PAG, peri-aqueductal gray; SNc, substantia nigra pars compacta; SNr, substantia nigra pars reticulata; Am, amygdala; LGN, lateral geniculate nucleus; OFC, orbitofrontal cortex. Note that many of the structures cannot actually be seen in a midsagittal slice; the indications serve only to provide a vague approximation, not an actual anatomical representation.

brain, more or less indicating the whereabouts of the neural structures most often investigated in relation to decision making, as well as some more basic structures for visual processing—following the bias in contemporary neuroscience toward studying decision making using experimental paradigms that focus on visual processing. The Wurtz-like methods of recording from single neurons allow us to insert an electrode in any of these structures to zoom in on one aspect of the spatiotemporal neural pattern in the brain while the subject performs a task. Data from a single neuron here or there might correlate or not with this or that behavioral process, and when we do obtain a significant correlation, this is only the beginning. Next, we need to consider how the neuron got activated and where the signal will go—which neurons it interacts with, what kinds of neurotransmitters and receptors are involved, and so on.

As for the influence of rewards on neural circuits, the beginning of the investigation was made by Wolfram Schultz and colleagues (Hollerman & Schultz, 1998; Schultz, Dayan, & Montague, 1997; Waelti, Dickinson, & Schultz, 2001). Other investigators had already observed single neurons in the basal ganglia (Hikosaka, Sakamoto, & Usui, 1989) and the prefrontal cortex (Watanabe, 1996) that showed systematic changes in the activity level during a waiting period as a function of whether the monkey could expect to receive a reward. Schultz and colleagues, however, were recording from a different type of neuron, whose activity pattern looked suspiciously like what psychologists had conjectured to be needed for learning.

Rescorla and Wagner (1972), as well as Dickinson (1980), proposed that to acquire the association of a stimulus and a reinforcer (e.g., a reward), the subject would need more than the mere observation that the two follow one another in time. Some discrepancy or "error signal" should be in order as well, both when an unexpected reinforcer is given and when an expected reinforcer fails to be delivered. The two types of discrepancy or error would be crucial for learning. The "positive" error of an unexpected reinforcer enables a new association between a stimulus and a reinforcer. The "negative" error of an expected but omitted reinforcer would lead the subject to reduce or eventually even lose a previous association between stimulus and reinforcer (in a process known as "extinction").

Schultz and colleagues (Hollerman & Schultz, 1998; Schultz, Dayan, & Montague, 1997; Waelti, Dickinson, & Schultz, 2001) showed that the activity of midbrain dopamine neurons in monkeys corresponded very well with the hypothesized error signals in the case of positive reinforcers (rewards). These midbrain neurons have their soma in the substantia nigra pars compacta or the neighboring ventral tegmental area, and they typically have a spontaneous firing rate of around six to eight spikes per second—spikes that lead to the release of the neurotransmitter dopamine. When the monkey suddenly receives a drop of fruit juice, the dopamine neurons respond with a brief burst of activity, spiking a few extra spikes.

When the fruit juice consistently follows exactly one second after the presentation of a colored picture on the screen in front of the monkey, the dopamine neurons adapt, and after just a few trials, they respond with a burst of activity to the visual stimulus ("the predictor"), but not the fully predicted reward. Conversely, the dopamine neurons respond with a pause of activity, that is, a drop below baseline, spiking not a single spike for negative prediction errors, shortly after the predicted reward should have been presented but was omitted.

The work by Schultz and colleagues on the "dopamine prediction error" must be among the best acknowledged and most often cited anywhere in neuroscience and has inspired several lines of investigation in different directions. The first studies employed classical or Pavlovian conditioning procedures (requiring no instrumental behavior), but similar dopamine activity was obtained with operant conditioning (Kawagoe, Takikawa, & Hikosaka, 2004; Satoh et al., 2003—see also Schultz, Apicella, & Ljungberg, 1993, right before the rise of the "dopamine prediction error" concept). Other studies have shown the dopamine signals to be analog rather than digital, influenced not simply by the presence or absence of prediction error but also by parametric aspects, such as the size and the probability of reward (Tobler, Fiorillo, & Schultz, 2005), the recent history of rewarded trials (Nakahara et al., 2004), and the delay until reward delivery (Fiorillo, Newsome, & Schultz, 2008; Kobayashi & Schultz, 2008).

Other species also show dopamine activity relating to reward prediction. Dopamine neurons in rats signal both positive and negative prediction errors (Roesch, Calu, & Schoenbaum, 2007), though the positive prediction errors seem to coexist with increased activity to fully predicted rewards (Pan et al., 2005). In humans, the data remain scarce, as fMRI studies are usually not tailored to pick up signals from the brainstem structures where the dopamine neurons live, but Jonathan Cohen's group (D'Ardenne et al., 2008) specifically targeted the region and concluded that the activity pattern did not look exactly like the prototypical dopamine prediction signals. For primary rewards (like fruit juice) as well as monetary rewards the brainstem structures showed positive but not negative prediction errors. Instead, ventral striatal regions—a major target zone for dopaminergic projections, particularly from the ventral tegmental area—showed both types of prediction error.

There are further cracks in the story. Peter Redgrave and colleagues have forcefully argued that dopamine signals, occurring as early as seventy to a hundred milliseconds after stimulus onset, may be too fast to really compute the prediction errors hypothesized (Redgrave & Gurney, 2006; Redgrave, Prescott, & Gurney, 1999). The argument resounded all the more vigorously following a particularly elegant experiment in which Redgrave and his team used so-called "disinhibition" procedures in anesthetized rats (Dommett et al., 2005). Under anesthesia, the superior colliculus does not respond to visual stimuli, but the sensitivity can be restored temporarily by injecting bicuculline, a pharmacological agent that blocks a type of gamma-aminobutyric acid (GABA) receptor. GABA is the major inhibitory neurotransmitter, and so the effect of bicuculline is to inhibit inhibition, a double negative that opens the gate to visual information. With the gate open in superior colliculus, dopamine neurons became visually responsive. Yet restoration of visual signals in primary visual cortex did not translate into any response in dopamine neurons. Dopamine neurons, then, probably get most of their visual input directly from superior colliculus—input that would be fast indeed but rather coarse and of little use in discriminating images associated with reward from those that are associated with something less.

Of course, these notes and reservations with respect to the hypothesis of dopamine prediction error do not diminish the massive achievement of Wolfram Schultz in taking neurophysiology straight to the domain of computational theory. As stated at the outset, in looking for a place in the skull for all we know, there had to be a beginning, any beginning, a point of departure to chart entire circuits. Wurtz might have fine-tuned the paradigm of single-unit recording and shown it to be compatible with the investigation of decision making and all the elusive happenings inside the brain, but Schultz introduced the precise language of computation and neural modeling.

Perhaps the activity of dopamine neurons does not in and of itself represent the whole of reward prediction error, but it certainly correlates with it or contributes to it in some way. It will be the job of Schultz, Redgrave, and all of us to figure out in perennial future research exactly how the dopamine activity plays its part in knowing about the rewards that come with given contexts. We need to consider how dopamine activity interacts with other neural processes, where it comes from, and where it goes to. Also the concept of reward prediction error will benefit from refinement and should be considered in relation with other mental functions, most notably in the very emotional domain that Skinner had endeavored to remove from the original reinforcement picture as it was offered by Thorndike.

Even Skinner would not deny that pleasure and satisfaction really do exist—he just did not know how to embed them in his variety of formal theory. The question will certainly come back though, sooner or later. As soon as formal theory makes contact with neurophysiology, and neural activity seems to code some aspect of a cognitive process, emotion lurks behind the open door. With respect to reward prediction error, for instance, there would be nothing to signal at the moment when a fully predicted reward is delivered, and indeed, empirical fact shows there usually to be no dopamine pulse (if we briefly make abstraction of the exception noted by Pan et al., 2005). What does the lack of activity at this time imply for the emotions, for the feelings of happiness, pleasure, or satisfaction, which we all know do float around in the neighborhood of rewards? Does a fully predicted reward fail to make us happy, and does the pleasure shift, along with the dopamine pulse, to the moment of positive reward prediction error?

Looking at my two-year-old son eating yogurt a few days ago, I would indeed think the happiness resides at least partially in the prediction and sometimes not at all in the consumption. While swallowing down the last spoonful of his own portion, he already started staring at my cup. He broke into a smile as soon as he realized my spoon was headed to no other mouth than his, and then he got cranky again a few seconds later, with the yogurt still on his tongue. He wanted more (he often sings the Song of More). Where does the dopamine activity go, and where the pleasure, the arousal, and so forth? How confident are we that dopamine is not simply an excitement drug, or an arousal signal that alerts the entire organism to something concurrent that should be processed attentively, stored in memory, or otherwise given priority in decision making?

The full story must be more complicated, of necessity, but we will only be ready to receive it once we are willing to include the dynamics of emotion and conscious experience as parameters along with the behavioral and the neural. One way to get at least a little bit closer to this purpose would be by consistently charting the physiological underpinnings of emotion, in the form of arousal measures, along with behavioral performance in terms of accuracy and response time. Perhaps I should plead guilty to laziness on this point, examining my son's behavior and physiology in great detail using the intuitive and surprisingly powerful means of detailed observation but recording only a paucity of data from my subjects in experiments, usually only saccadic eye movements or button presses, no blood pressure, heart rate, galvanic skin response, pupil dilation, and so on. How would the activity of dopamine neurons correspond with such physiological measures? I am afraid no one has yet bothered to check.

Another thing that has always intrigued me about the story of reward prediction error is how the signals seem to be passive and waiting until suddenly something good comes along for the fireworks to begin. It does not jibe with my idea of motivation and reward-oriented behavior. I would like to think of the possibility of reward as something that engages search behavior, an active form of looking forward. Returning to the Skinner quote about search behavior (and figure 2.1 with the hunter–gatherer model), the reward prediction error would be about learning, about storing past successes in memory and associating rewards with the statistics of the environment, but not about driving the behavior or energizing the actual search. In dangerously phenomenological terms, I thought the story was too much about liking (what you have) and not enough about wanting (what you do not have).

Therefore, I went looking for neural activity that was of the more desirous type, prospective, and anticipatory. Not I, but we, of Okihide Hikosaka's laboratory, and to put the chronology in correct order, the search had begun already when I joined the laboratory in May 1998. We would investigate eye movements in the dark and how they look for what there is to see. We would study their underlying architecture, the rules and habits, the laws of irrigation, and what reward is and what good is. It sounded like some kind of incorrigible music to my ears (to be precise, the third album by OMD—a British "synth pop" group that brutally dominated the airwaves in supermarkets and fast-food restaurants in the early 1980s). (Not that I did not like it then.)

Ocular Movements in the Dark/Architecture and Morality

In a prototypical Wurtz-like paradigm, the monkey would sit in the dark, make eye movements, have an electrode sitting in its brain, and sometimes receive a drop of fruit juice or Pocari Sweat (the most unlikely true name for a Japanese sports drink). Hikosaka and colleagues had already studied neural activity in the monkey caudate nucleus relating to reward since the 1980s (Hikosaka, Sakamoto, & Usui, 1989; Kawagoe, Takikawa, & Hikosaka, 1998) and had established that caudate neurons show increased activity following a visual target that is associated with reward in a "memory-guided" saccade task. In this task, the monkey has to make an eye movement in the dark to a position where a visual target was briefly flashed more than a second before.

Neurons in the caudate nucleus (part of the dorsal striatum, the input structure of the basal ganglia) seemed to have access to reward information all right. This made perfect sense, as the basal ganglia were thought to be an important brain area for the integration of context and action, or in one word: motivation (Mogenson, Jones, & Yim, 1980). Basal ganglia deficits affect the voluntary control of behavior, as in Parkinson's disease (Brown et al., 2006) and can lead, specifically following caudate lesions, to "abulia," or a lack of will (Bhatia & Marsden, 1994; Caplan et al., 1990). Anatomically speaking, the basal ganglia do seem to be perfectly placed to have something to do with motivation, or attaching reward information to action plans. The basal ganglia get direct excitatory input from dopamine neurons, as well as from much of cortex, and send inhibitory output to thalamus (and then back to cortex) and superior colliculus—two pathways for indirect motor control (Groves, Linder, & Young, 1994; Smith & Bolam, 1990).

In the course of experimentation, Hikosaka and colleagues observed a more intriguing species of neuron—a type they had not bargained for. Yet, the activity looked so strikingly peculiar and idiosyncratic that it just had to mean something. This type of neuron showed vigorous anticipatory activity *before* the appearance of the visual target, that is, at a point in time when the monkey could have no clue about where the next eye movement should go. The anticipatory activity, moreover, was clearly context dependent—it happened in some blocks of trials, not in others. Careful examination prompted the conclusion that it must somehow depend on the association between a spatial position and the availability of reward (Takikawa, Kawagoe, & Hikosaka, 2002). The neurons seemed to prefer one specific combination, and so they would fire vigorously in anticipation of visual targets during blocks of trials with the "coded" position–reward association, whereas they would be virtually silent during other blocks of trials. The grand hypothesis (uttered only over lunch in the cafeteria of Juntendo University in Tokyo) was that "caudate neurons have desire fields," analogous to receptive or mnemonic fields but prospective instead of retrospective. However, Hikosaka was quick to admit that the label sounded premature, a leap too far into the unknown.

The problem was that the experimental paradigm did not really allow us to disambiguate the activity in any clear functional terms. Of course, the paradigm had not been tailored to investigate the anticipatory activity in the first place. Now the visual target, its incentive value, and the eye movement were all centered on the same position in space, and so the neural activity could be related to perceptual, motivational, or motor processes—or any combination of these three. Moreover, the memory-guided task, with a delay of a second or more between the visual target presentation and the eye movement, prevented a detailed behavioral analysis of response time. The paradigm included a randomly interleaved set of reward trials and no-reward trials (or large-reward trials and small-reward trials), but the behavioral performance did not seem to be affected very clearly by the reward factor. There were no significant differences in response time, and though the performance was more accurate in reward trials than no-reward trials, there were usually not enough error trials for a proper statistical comparison of the neural activity in correct versus error trials.

Seeing as I was so adamant about knowing exactly what needed to be done, Hikosaka suggested I should do the right thing and take charge of the necessary follow-up studies. In one study, we manipulated the relation between sensory information and incentive value (Lauwereyns et al., 2002a). We continued to use the same position–reward conditions, but we added color–reward conditions, in which the visual target and the eye movement were still centered on the same position in space but the incentive value varied with the color of the visual target, not the position. Again the caudate neurons showed anticipatory activity that depended on the reward context, but this time the modulation was governed by color as much as by position—some neurons fired more strongly for a specific color–reward combination, and other neurons "preferred" any color–reward combination over any position–reward combination.

Looking more closely at the data, we noticed that virtually all of the caudate neurons with anticipatory activity also showed a discriminative visual response. Applying the logic of signal detection theory and the search for neural signatures, as outlined in chapter 1, we went on to examine the scaling properties of the caudate neurons and obtained beautiful evidence of an additive mechanism—the visual input looked to be riding the wave of anticipation. The data suggested that the anticipatory activity functioned as a bias mechanism that could bring the neural system closer to the threshold of a perceptual decision. It was one very plausible brain mechanism for favoring the hypothesis that the visual target is associated with reward—the functional equivalent of wishful seeing.

We now had some evidence that the anticipatory activity could bias the visual processing, but the link to wishful seeing was still very tenuous, or merely implicit. To complete the picture, we needed to adapt the paradigm to enable response time analysis. By removing the delay between the visual target and the eye movement, we finally did obtain crisp response time distributions that showed a substantial effect between reward and no-reward trials (Lauwereyns et al., 2002b). Again, we encountered neurons with context-dependent anticipatory activity, spiking vigorously before the presentation of the visual target only in blocks of trials with the "coded" position–reward association. With the new data, however, we were able to confirm that the anticipatory activity covaried with differences in response time.

Figure 2.3 presents exemplar activity of a single neuron in caudate nucleus during the visually guided eye movement task with asymmetrical reward. The data are shown trial by trial but in a reordered sequence—not in the order as experienced by the monkey but ranked by the response time from the trial with the fastest response (top) to the trial with the slowest response (bottom).

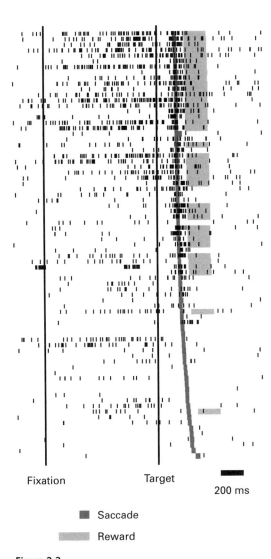

Figure 2.3
Data from a single neuron in monkey caudate nucleus (reported in a different format in Lauwereyns et al., 2002b). Each line represents one contralateral saccadic eye movement trial in an experiment with an asymmetrical reward schedule; each tick represents a spike fired by the neuron. The trials are rank ordered by the response time, from short (top) to long (bottom). Reward trials are marked by a long gray bar. The left vertical line indicates the moment of fixation onset; the right vertical line indicates the moment of target onset. Here, response time tends to be shorter for reward trials than for no-reward trials. This advantage in response time correlates with the neuron's anticipatory activity.

Only data from trials with a contraversive eye movement are shown, that is, trials in which the eye movement was directed to the contralateral hemifield. For this particular neuron, recorded from the left caudate nucleus, the trials included in figure 2.3 are those with an eye movement to the right hemifield. To be sure, the monkey performed an equal number of ipsiversive and contraversive eye movements and an equal number of reward and no-reward trials, all randomly interleaved (trials with ipsiversive eye movements are not included in figure 2.3, as they showed no relation between neural activity and response time).

Just glancing at the figure, it is obvious that the anticipatory activity, between the moment of fixation onset and the target appearance one second later, strongly correlates with the response time for a contralateral eye movement. For trials with a short response time, there tend to be markedly more spikes in advance of the visual target than for trials with a long response time. Another striking feature of the anticipatory activity is how it ramps up toward the target onset—a characteristic that may well be one of the best telltale signs of prospective coding (Rainer, Rao, & Miller, 1999). The evidence, taken together with the other caudate data, pointed strongly in the direction of a reward-oriented bias mechanism, making the neural system particularly susceptible to false alarms of the kind that arise when we believe wishes to have come true, when they may very well have not.

One disappointment, however, was that the analyses of the response time data according to the LATER model produced inconclusive results (and so the Carpenter papers had to remain conspicuously absent from the discussion in Lauwereyns et al., 2002b). Since then, Nakahara, Nakamura, and Hikosaka (2006) had better luck with data from a similar behavioral paradigm, but it came at the cost of two extra parameters to the LATER model. Alternatively, doing away with the deadline for eye movement initiation, and effectively outlawing the concept of error, it is possible to get response time data from rats that correspond well with the original LATER model (Lauwereyns & Wisnewski, 2006). In this adaptation of the paradigm, the rats are trained to make operant nose-poke responses, that is, to poke their nose into a target hole illuminated by a light-emitting diode. In essence, this reflects an orienting response, performed not with the eye alone but with the entire body. As the rat pokes its nose into the correct hole, it breaks an infrared beam of light, which gives a very precise, and entirely noninvasive, measurement of response time—definitely the paradigm of the future for me. Using the visually guided nose-poke paradigm with an asymmetrical reward schedule, we were able to confirm that most of the systematic variation due to the reward factor happened exactly in accordance with the hypothesis of a bias mechanism in LATER language, saying that the distance between the starting point and the threshold for the decision process got reduced.

Now I would wish to see dorsal striatal neurons in the nose-poking rat doing what the homologue caudate neurons do in the eye-moving monkey, so that there can be some proper triangulation from response times and neural activity to decision making. The state of the art, then, is very much in mid flight. While we have a few unmistakable clues that enable us to start painting a fairly detailed picture of reward bias in a visual orienting paradigm, there remains quite a bit of work to be done before we can confidently believe we have explained how it works—unless, of course, we use the word "explained" as in Dennett's (1991) *Consciousness Explained*. Nevertheless, data-driven speculation has its place in science, and perhaps even more so in a book-length monograph than in a succinct review article. Figure 2.4 gives my best and leanest approximation of a neural circuit for eye movement control that incorporates reward-oriented bias (heavily based on the anatomical circuits as outlined in Hikosaka, Takikawa, & Kawagoe, 2000).

The superior colliculus gets top billing as the ultimate neural structure for the initiation of eye movements. From the superior colliculus, motor commands go downstream to brainstem nuclei that translate the target coordinates of an eye movement into the appropriate push and pull of the six muscles that are attached to the eyeball. At the level of superior colliculus, we are still in contralateral territory—neurons in the left hemisphere code for eye movement vectors aimed at the right half of visual space, and vice versa. The signals in superior colliculus are derived from two broad streams of input: a stream of excitatory inputs from cortex (especially frontal eye field, as shown in figure 2.4, but also lateral intraparietal cortex and other structures, not included here to prevent clutter in the minimalist scheme) and inhibitory inputs from the basal ganglia, particularly substantia nigra pars reticulata. The excitatory cortical inputs (causing superior colliculus neurons to fire action potentials) use the neurotransmitter glutamate, whereas the inhibitory reticulata inputs (suppressing collicular activity) work with GABA.

An important note here is that the reticulata neurons, by default, show a very high baseline of spontaneous activity, some neurons spiking as many as fifty or sixty spikes per second or even more. This is very different from the six or seven spikes that dopamine neurons produce, for instance, and listening to the action potentials, it sounds like we go from torrential rain to a mere trickle when we move the electrode from pars reticulata to pars compacta. The high baseline of inhibitory activity by substantia nigra pars reticulata acts like a dam, preventing superior colliculus from getting easily excited about anything. This means that superior colliculus neurons can only initiate an eye

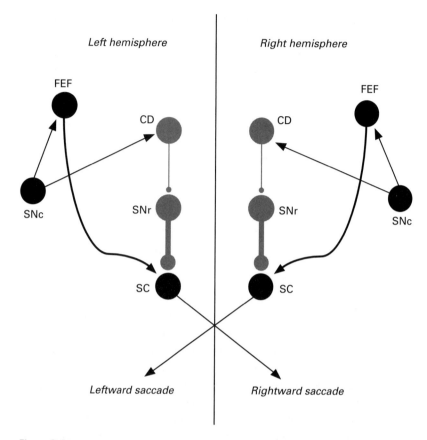

Figure 2.4
Schematic presentation of a neural circuit for eye movement control. Black arrows indicate excitatory connections; gray arrows indicate inhibitory connections. The thickness of the arrows represents the strength of activity. The indicated structures are a subset of those in figure 2.2: FEF, frontal eye field; CD, caudate nucleus; SC, superior colliculus; SNr, substantia nigra pars reticulata; SNc, substantia nigra pars compacta. The superior colliculus integrates excitatory inputs from cortex (e.g., FEF) with inhibitory inputs from the basal ganglia (particularly SNr) to derive motor commands that are then sent downstream to brainstem structures for eye movement generation.

movement when the driving cortical activity is strong enough to overcome the suppressive reticulata activity.

In essence, we can think of the decision-making process for an eye movement as a competition among several inputs to the superior colliculi in both hemispheres. Though not explicitly shown in the circuit, there also exist lateral connections among the structures in the different hemispheres—from the left superior colliculus to the right superior colliculus, and vice versa, or even from the left substantia nigra pars reticulata to the right superior colliculus, and vice versa. We can think of the competition as a kind of voting contest in which the brain chooses among different superior collicus candidates, representing different eye movement vectors. Nonpreferred candidates are attacked with vicious slander (a lot of GABA), while the bad-mouthing is strategically withheld for more acceptable candidates. In the meantime, vigorous endorsements for the favorite candidate hope to create the decisive push. The superior colliculus will actually initiate an eye movement as soon as one vector manages to tip the balance, or reaches the threshold for activation.

Some of the voting mechanisms are quite complex, involving publicists that pull the strings of voters, or transactions that occur in the shadows. For instance, the inhibitory activity from substantia nigra pars reticulata can, in turn, be inhibited by the GABAergic projection from caudate nucleus, an input structure of the basal ganglia. This route from caudate nucleus to substantia nigra pars reticulata is part of the direct pathway in the basal ganglia; there is also an indirect pathway from the caudate nucleus to the substantia nigra pars reticulata, which takes a detour via the globus pallidus external segment and the subthalamic nucleus and actually leads to the opposite effect, amplifying the suppressive reticulata activity. Perhaps the most sensible idea is that the pull from caudate via the direct pathway serves to selectively remove reticulata inhibition for target vectors, whereas the push via the indirect pathway would strengthen reticulata inhibition for all the other vectors—this is just another bit of speculation that has to be checked empirically. For simplicity, I draw only the direct pathway in figure 2.4. In any case, the caudate nucleus regulates the inhibitory activity from substantia nigra pars reticulata and so has the power to remove the dam for cortically inspired eye movement instructions in the superior colliculus.

Now to put the icing on the cake or (in a more pessimistic analogy) to introduce the role of money and lobbyists in politics, we have dopaminergic projections from substantia nigra pars compacta to implement the reward factor in cortical structures as well as in the caudate nucleus. To be sure, the caudate nucleus also gets input (not shown in figure 2.4) from frontal eye field and other cortical structures, so another way to summarize the whole would be by saying that cortical structures control superior colliculus by pushing via a direct route and pulling via the basal ganglia (with another push-and-pull mechanism embedded there)—keeping in mind that the cortical input to the basal ganglia is modulated by dopamine (see Reynolds, Hyland, & Wickens, 2001, for a particularly impressive demonstration). The activity in caudate nucleus, then, should not be thought of as a mere copy of what the cortex says (Ding & Hikosaka, 2006; Pasupathy & Miller, 2005).

Taking the neural circuit in figure 2.4 as the basic setting when everything is equal and neutral, we can now move on to figure 2.5 for a look at what happens in the circuit when the reward context favors one eye movement over another—here, an eye movement to the right. The most exciting action is in the left hemisphere, so we can take the liberty of forgetting about the right hemisphere for a minute. With dopaminergic input, the caudate nucleus in the left hemisphere shows enhanced anticipatory activity, which inhibits the suppressive activity from superior colliculus and lifts the blockage specifically for rightward eye movements. Studies of activity in substantia nigra pars reticulata (Sato & Hikosaka, 2002) and the superior colliculus (Ikeda & Hikosaka, 2003) confirm that the biased anticipatory activity behaves in exactly the way we would expect if the neural circuit does what we think it does. In the end, under the regime of biased anticipatory activity, the superior colliculus neurons are more easily pushed into action with the excitatory input from the frontal eye field.

Put differently, the anticipatory activity in the basal ganglia effectively produces a change in the baseline activity of the superior colliculus. As a consequence, the relevant neurons are already closer to the threshold for initiating an eye movement when the action brings something desirable than when it gives something less (see Lo & Wang, 2006, for a formal implementation). This proposal perfectly resonates with the bias mechanism in the LATER model, implying a change to the starting point for the decision process, or a shorter distance to travel on the way to the threshold. Perhaps I am reading too much into the neural circuit, but I think, in hindsight, that the double inhibitory projection via the basal ganglia—a double negative to create a positive that does not interact with other inputs—is really the optimal way to achieve additive scaling in the way postulated in chapter 1 for shifts of signal and noise distributions, using the framework of signal detection theory. In any case, I do have the impression that the pieces of the puzzle are starting to fall in place and that the neural circuit and its dynamics as shown in figures 2.4 and 2.5 agree with the proposals of anticipatory processing and additive scaling, all in the business of reducing the distance that the decision process has to cover on its way to the threshold.

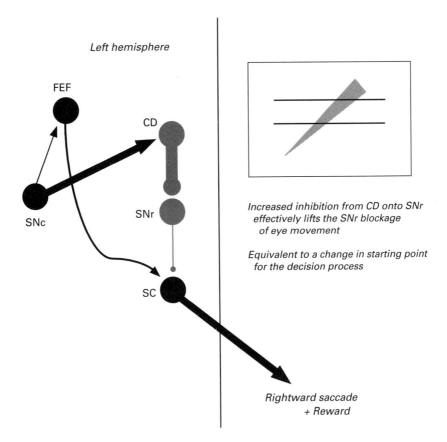

Figure 2.5
Effects of bias in a neural circuit for eye movement control. The schematic presentation follows the format of figure 2.4, with the following neural structures: FEF, frontal eye field; CD, caudate nucleus; SC, superior colliculus; SNr, substantia nigra pars reticulata; SNc, substantia nigra pars compacta. Now, rightward saccades are associated with reward. The reinforcement schedule implies no changes for the right hemisphere (not shown), but in the left hemisphere, it leads to dopamine input from SNc. This enables heightened anticipatory activity in CD, which suppresses the default inhibition from SNr onto SC. As a result, FEF can more easily push SC to initiate a saccade. This mechanism corresponds to a shift in starting point according to the LATER (linear approach to threshold with ergodic rate) model (inset figure to the right).

There is, alas (or luckily, for those of us who love a good mystery), one thing that remains rather enigmatic in the neural circuit. What exactly does the dopamine input do here? How can we get from a reward prediction error that occurs *after* stimulus onset to anticipatory activity *before* stimulus onset? We know for sure that the dopamine neurons have a different activity profile, indifferent to the pretarget period, so we can rule out any notion of concurrent modulation. Another, perhaps more neutral way to phrase the problem is by asking how the caudate neurons get stimulated into ramping up their activity toward target onset. We could defer the issue and say the activity is simply fed down from the frontal eye field, and this may be true to some extent (Ding & Hikosaka, 2006). It may also be true that the supplementary eye field has something to do with it (Coe et al., 2002). But then how do the frontal and supplementary eye fields get started with their anticipation?

Dopamine projects to the frontal and supplementary eye fields as well as to the caudate nucleus, and so it does seem to be well placed for some kind of influence, whichever way we turn. Then what, if anything, could dopamine do? Given that it does not engage in any concurrent modulation, there remain just two possibilities. Either dopamine does nothing at all for the reward-oriented bias mechanism, or it works across time, from trial t to trial $t + 1$. With effects over time, we return to the topic of learning, and maybe this is the true turf for dopamine after all, more than simply arousal or something hedonic in the processing of reward information. By this account, dopamine would be responsible for the dynamic control of synaptic weights in the neural circuit. This brings us to the esoteric world of long-term potentiation, and its cousin, long-term depression, by which synapses between neurons are strengthened or weakened. It would involve intracellular mechanisms and structural changes, such as the growth or demise of dendritic spines or an increase or decrease in the concentration of a particular receptor type.

Evidence in favor of this kind of dopamine-dependent synaptic plasticity comes from a brilliant study by Reynolds, Hyland, and Wickens (2001), one that I already mentioned a few paragraphs ago. (I feel compelled to cite it wherever I can—it had a large enough impact on my life, bringing me quite literally to New Zealand.) The three kiwis combined intracellular recording in dorsal striatal neurons with stimulation in substantia nigra pars compacta (providing dopaminergic input to the striatum). They used stimulation sequences in substantia nigra pars compacta that matched with those the rat had learned to press for in an intracranial self-stimulation (ICSS) paradigm, during a training phase before the intracellular recording. The recording, though, had to take place under anesthesia—the fact that it was done in a whole, live brain was already a giant step up from working with the usual expired and completely deafferentated bit in a slice preparation.

The ICCS paradigm was modeled after a classic study by Olds and Milner (1954), which had shown that rats will learn to work for electrical stimulation through an electrode inserted in their own brain, as long as the stimulation reaches an area that can send the right kind of reinforcement signal to the rest of the brain. Stimulation in the substantia nigra pars compacta must "feel good" if we indulge (once again, very briefly) in phenomenological talk. However, regardless of putative hedonics, Reynolds, Hyland, and Wickens (2001) were able to show that the stimulation in substantia nigra pars compacta affected the slope and the peak amplitude of the postsynaptic potentials in dorsal striatal neurons (which had separately been shown to respond to cortical input). Moreover, the ICSS-like stimulation effects could be blocked by injecting the rat with a dopamine receptor antagonist. There remained only one, beautifully simple conclusion. The dopamine input from substantia nigra pars compacta strengthens the synapses between cortical and dorsal striatal neurons.

Thanks to dopamine, every glutamatergic pulse coming in to the dorsal striatum would be more likely to elicit a decent response from the GABAergic neurons, as if the signal reception is now more efficient, the synapse tighter. Applied to the reward-oriented bias, it means that a dopamine signal in a previous trial would lead to an amplification of the caudate response to cortical input in the next trial. Here, I can happily refer to Nakamura and Hikosaka (2006a), who established that intracranial electrical stimulation (not by the self but by the experimenter) in the caudate reinforces future eye movements if applied right *after* the previous eye movement. The data encourage us to think further in terms of plastic changes over time. It does seem to be the most promising route to a proper understanding of how reward-oriented bias develops.

But how does the anticipatory activity actually manage to ramp up, or grow in strength? Perhaps there would be some kind of amplification that reverberates through a feedback loop, from cortex to the basal ganglia, and back to the cortex, cycling through the local circuit a few times before heading off to the superior colliculus. In figure 2.4, I emphasized the projection from substantia nigra pars reticulata to superior colliculus, but in fact the majority (at least two thirds) of the output from reticulata goes to the thalamus, which in turn, sends its signals to several cortical regions. Thus, we can imagine a scenario in which there is an initial random or somehow sensory-driven spark in cortex. The spark gets amplified in the caudate and travels back to cortex, only to enter caudate again, get even more amplified, and keep echoing louder and louder, the activity ramping up toward the moment of target onset. All the while, dopamine would be silent, lying on its back, contemplating a job well done. The bridges were built.

Quite a bit of empirical work remains to be done before we can carve this figure of the neural circuit in stone. At present, the data are particularly blurry for the human brain. There has been a massive surge of fMRI research on reward-related information processing but relatively few studies focused on anticipatory activity *before* the presentation of reward-predictive information—notable exceptions are Critchley, Mathias, and Dolan (2001), who focused on "anticipatory arousal," and Bjork and Hommer (2007), who as of yet probably came closest to an instrumental paradigm that would provide an opportunity to compare anticipatory activity with reward-dependent effects in behavioral performance. Sooner or later, someone will have to try to correlate the dynamics of BOLD responses with behavioral measures of bias, such as those offered by Liston and Stone (2008), in a task with an asymmetrical reward schedule.

For Wurtz-like studies, there is plenty of work ahead as well. We might wish to correlate the amplitude of a dopamine pulse in a previous trial with the development of anticipatory caudate activity in future trials. Or we would be wise to temporarily block dopamine transmission and check the consequences for reward-oriented bias. Actually, the truly inevitable Hikosaka, working with Nakamura, has already collected preliminary data of this sort, looking at the effects on response times after injecting pharmacological agents that interact with dopamine in the caudate (Nakamura & Hikosaka, 2006b).

Another aspect of the anticipatory activity that needs to be looked at more closely is how it depends on contextual information that sets the stage, in the four dimensions of time and space, for reward-predictive events. In the eye movement task with asymmetrical reward, the trial started with the presentation of a fixation point at the middle of the screen (this moment was indicated by the left vertical line in figure 2.3). The monkey had to look at the fixation point and wait exactly one second for the target to appear either to the left or to the right. There was ample opportunity to start anticipating in a strategic fashion, aimed at a predictable event. We can think of the fixation onset as the predictor of a reward predictor, or a first signal that alerts the monkey to the impending arrival of some interesting information. Scrutinizing the spike data in figure 2.3, we notice that the anticipatory activity usually starts at about three hundred milliseconds after the onset of the fixation point, sometimes a bit later, but never before the first cue arrives on the scene. Before fixation onset, there is only the random rumble of baseline, a meaningless bit of spontaneous activity.

The prospective mechanism here is quite clearly focused on a specific moment in time, relying on the probability structure of events in the task. Given that the target will appear one second after fixation onset, it makes perfect sense for the neural system to activate a bias mechanism that would enable it to respond fast to a target associated with reward. Wishful seeing is most relevant in the twilight, when we know there is a good chance that there could be something. We might even think of it as a recursive bias operation, in which the bias mechanism is biased to occur when the prior probability of a relevant event is high.

In any case, the predictability of target onset does have a large impact on the development of anticipatory activity, as shown by Hikosaka and colleagues in an early study of caudate activity (Hikosaka, Sakamoto, & Usui, 1989). When the delay between fixation point and target was fixed at one second, the caudate neurons neatly ramped up their activity. For variable delays, the caudate neurons only managed to erratically spike somewhat more often than during intertrial intervals.

The moral of the story must be that anticipation, wishful seeing, and the various varieties of looking forward will be all the better defined when the space-time coordinates of the thing in question are as detailed and determined as can be. It works most efficiently following a clear trigger, an initial "get ready," or the first striking of a chord, more to get the audience to stop chatting and start listening than to tune the violin.

After Each Stroke the Vibrations

We might be very eager to welcome a long-awaited friend to our home, but our neurons will not know when to get active unless we have knowledge of our friend's train schedule (as well as cultural information about the likelihood of trains being on time—not the same for Japan and New Zealand). With less precise coordinates, our anticipations may wax and wane, oscillate in the background of our thoughts, and temporarily lapse into nonexistence, as we get engrossed in reading about the self-created misfortunes of marijuana-smoking Russian sumo wrestlers. The bias mechanism would suddenly wake up, though, and fire up again with a knock on the door, only to bring no small amount of disappointment when we recognize our all too cheerful neighbor, selling chocolate bars for the local baseball (or cricket) team. Yet, independent of the negative emotions upon being confronted with the evidence of a false alarm, our anticipations might linger a little while longer, the bias mechanism still ringing a bit. Or as David Hume (2002, p. 282) put it in his inimitable style:

Now if we consider the human mind, we shall find, that with regard to the passions, 'tis not of the nature of a wind-instrument of music, which in running over all the notes immediately loses the sound after the breath ceases; but rather resembles a string-instrument, where after each stroke the vibrations still retain some sound, which gradually and insensibly decays.

The eighteenth-century Scotsman might in the first place have thought of retrospective cognitions when he composed this gorgeous sentence, but I think it works in either direction, with or against the arrow of time. The fact is that the life of neural activity in the brain, somewhere in the dark between sensory input and motor output, exhibits growth and decay as readily as any other biological phenomenon. And even the sensory input and motor output are not finite borders around the life of neural activity, as anticipations are born before the physical variations in light or sound, and other codes might ruminate about whether a certain decision was a good one, long after the eye has moved or the thumb has pressed a button.

The time course of neural activity is certainly one of its most useful identifying characteristics, convenient for comparison when we look for commonalities and correspondences among different brain structures. Rainer, Rao, and Miller (1999) went one step further in suggesting that growth of activity indicates prospective coding, whereas decay would imply retrospective coding. I am not entirely sure whether the logic holds—for one thing, the level of activity should be compared against a baseline, so that a gradual drop of activity in, say, the substantia nigra pars reticulata can be recognized as a form of growth. However, I do believe provocative ideas such as these should be explored more often in contemporary neuroscience.

With respect to influences of reward on decision making, it will be obvious that not all of these are captured by anticipatory processing and changes to the starting point for a decision line. There must be other processes at play, showing different types of temporal dynamics. Shuler and Bear (2006), for instance, showed posttarget reverberations in the rat primary visual cortex that were tuned to the timing of reward. As laid out in chapter 1, we can also look for evidence of sensitivity effects as a function of reward. Might dopamine introduce a reward factor that works synergistically with perceptual codes, leading to multiplicative scaling, or the amplification of signals associated with reward relative to those that are not?

A number of Wurtz-like studies have already shown data to say "probably, yes." Especially for neurons in the dorsolateral prefrontal cortex, it is quite clear that the spatial information carried by the neural code is improved substantially when the visual targets are associated with a large reward (Kobayashi et al., 2002; Leon & Shadlen, 1999; Watanabe, 1996). Also in the frontal eye field (Ding & Hikosaka, 2006) and in the superior colliculus (Ikeda & Hikosaka, 2003) a sizeable proportion of the neural population shows a dramatically improved signal-to-noise ratio for processing visual targets that come with a large reward. In all of these studies, we can easily recognize the neural signature of sensitivity, with synergistic processing and multiplicative scaling in conjunction with enhanced behavioral performance for trials with a large reward.

Bringing the data to bear on the neural circuit for eye movement control, the reward-dependent sensitivity appears to travel predominately via the direct route from cortex to superior colliculus. Figure 2.6 shows how this works, in a format analogous to that of figure 2.5. Again we take the case that a rightward eye movement is associated with reward, and we concentrate on the brain half where most of the action takes place. With dopamine input, the signal from the frontal eye field and other cortical areas gets amplified and exerts a strong excitatory influence on the superior colliculus. Even with an unchanged baseline of suppressive activity from the substantia nigra pars reticulata, it would now become easier to tip the balance, outdo whatever happens in the superior colliculus in the other hemisphere, and drive an eye movement.

The proposed dynamics in the neural circuit for eye movement control would occur synergistically, in concert with incoming sensory information. The multiplicative scaling, producing an improved signal-to-noise ratio, effectively achieves the change to the gradient of the decision line as postulated for heightened sensitivity within the LATER model. And again I note how fitting it is that the multiplicative scaling should be based on the excitatory input from cortex, thinking that the convergence (and coincidence) of multiple excitatory inputs might very well lead to the emergence of nonlinear phenomena, in which the whole is more than the sum of its parts. In any case, the sensitivity mechanism would get eye movements to speed up for the more appetizing stimuli. James Stewart was right; it is a wonderful life after all, in accordance with the 1946 movie by Frank Capra. Together with the bias story in the previous section, we now have two mechanisms that can both influence behavioral performance as a function of reward.

There is nothing to prevent the sensitivity and bias mechanisms from peacefully coexisting. In fact, we already have in existence proof that the two are not mutually exclusive. Ikeda and Hikosaka (2003) encountered neurons in the superior colliculus that combine reward-oriented anticipatory activity and additive scaling with reward-oriented synergistic activity and multiplicative scaling. The real question is whether the bias and sensitivity mechanisms can exist in isolation. This is a question that I must return to in other chapters, and ask more broadly for bias versus sensitivity mechanisms in general, but here, concentrating on the neural underpinnings of happy prospects and wishful seeing, my hunch is that reward-oriented bias and sensitivity may be inseparable. I will quickly add that the matter should be investigated empirically, but

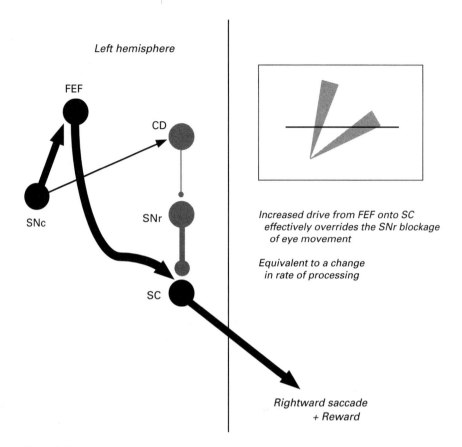

Figure 2.6
Effects of sensitivity in a neural circuit for eye movement control. The format and the neural structures are the same as in figures 2.4 and 2.5: FEF, frontal eye field; CD, caudate nucleus; SC, superior colliculus; SNr, substantia nigra pars reticulata; SNc, substantia nigra pars compacta. As in figure 2.5, only rightward saccades are associated with reward. Again we note the reinforcement schedule implies no changes for the right hemisphere (not shown), but in the left hemisphere, it leads to dopamine input from SNc. This boosts the activity in FEF, which creates a stronger push in SC to overcome the inhibitory input from SNr. This mechanism corresponds to a steeper gradient for the decision line according to the LATER (linear approach to threshold with ergodic rate) model (inset figure to the right).

if it is true that for both mechanisms the basic source of reward information comes from dopamine, the most likely answer will be that dopamine input to the one means dopamine input to the other as well.

Even so, it is not very hard to imagine that the two mechanisms could behave independently if the circumstances were right. The dopaminergic input to cortex, that is, the mesocortical pathway, originates from the ventral tegmental area, whereas the dopaminergic input to the caudate nucleus (or dorsal striatum), via the so-called nigrostriatal pathway, originates from substantia nigra pars compacta. Perhaps there are really two classes of dopamine neurons which can functionally be distinguished but have not yet been offered the right experimental opportunity to show off their differences. Another possibility is that the temporal dynamics imply functional differences. For the bias mechanism, I have suggested that we need to think in terms of influences over time and plastic changes. This approach may be relevant as well for the sensitivity mechanism, but here the dopamine prediction error is concurrent with the amplified visual response. Dopamine, then, might very well be an active ingredient in the synergistic mixture. Long-range and short-range dopamine influences might operate with different agendas. Nevertheless, the simplest assumption would be that reward-oriented bias and sensitivity naturally go together as two parallel mechanisms that employ different means to achieve the same goal, that is, giving more weight to options that imply reward.

To date, the best evidence for reward-oriented sensitivity has come from Wurtz-like studies. Relevant fMRI reports are beginning to emerge, explicitly connecting reward with neural correlates of improved performance, possibly due to sensitivity (Krawczyk, Gazzaley, & D'Esposito, 2007; Pleger et al., 2008). But to learn more about multiplicative scaling by reward, we can also take a look at the bulk of data on neural mechanisms of selective attention, produced by Wurtz-like studies as well as fMRI investigations and any other tool available in cognitive neuroscience. Maunsell (2004) correctly pointed out in a polemic piece for *Trends in Cognitive Sciences* that there are many similarities between the experimental paradigms used to study effects of reward expectation and those in service of characterizing selective attention. Yet, most researchers seem to follow only one of the two lines of investigation without much regard for the other. Studies that show multiplicative scaling effects as a function of attention in visual cortical areas such as V4 and MT have typically used reward in an operant conditioning paradigm to prioritize some visual stimuli over others (e.g., Luck et al., 1997; McAdams & Maunsell, 1999; Treue & Martínez Trujillo, 1999). Conversely, studies on reward expectation have required subjects to perform visual-discrimination tasks, employing the very same mental faculty of attention (e.g., Lauwereyns et al., 2002a; Platt & Glimcher, 1999; Roesch & Olson, 2003).

Maunsell's (2004) solution was to disentangle the investigations by manipulating factors such as stimulus similarity in the name of attention, and reward magnitude and probability in the name of reward. These are excellent suggestions, of course, but I think we would do one better if we can also include the measures of bias and sensitivity, examining additive versus multiplicative scaling and anticipatory versus synergistic processing. If we do so, we might see that the putative mechanisms of attention are multiplicative, synergistic, and aimed at reward, that is, they are simply another name for reward-oriented mechanisms of sensitivity.

In one sense, my call for a distinction between bias and sensitivity echoes an earlier one in the field of cognitive psychology when the concept of attention as proposed by Posner (1980) appeared ambiguous in terms of signal detection theory (see Downing, 1988, for the strongest indictment). Researchers have all too easily formulated their theories of attention with grand notions of facilitated visual processing, implying heightened sensitivity and improved signal-to-noise ratios, while their experiments left room for effects of bias as well as sensitivity. I think we can and should be more precise in our formulations. Attention is a useful, somewhat vague word, only to be used when we deliberately wish to hedge our propositions. For a proper investigation of how neural circuits weigh the options, attention may be less suitable than the pair of bias and sensitivity.

In the meantime, we can use the label of attention as a search term to locate studies that are relevant to the topic of reward-oriented bias and sensitivity, on the assumption that humans participate as subjects in experiments with some kind of reward in view, be it a few dollars more or the very desirable course credit, a sine qua non for graduation with a bachelor's degree in psychology. In these experiments, then, the subject's compliance with task requirements actually implies reward-oriented top–down control of visual processing. This affects sensitivity, say, in studies that show variations in visual activity as a function of task difficulty or the signal-to-noise ratio, complete with synergistic processing and multiplicative scaling (e.g., De Fockert et al., 2001; Liu, Larsson, & Carrasco, 2007). And it affects bias, for instance, in studies showing pretarget modulation, or anticipatory processing, possibly accompanied by additive scaling (see Kastner et al., 1999; Egner et al., 2008). Again, dopamine may be needed for all this to happen—some promising evidence exists already with humans in the scanner (Nagano-Saito et al., 2008). After drinking either a "nutritionally balanced mixture" or a rather ominous sounding "amino acid mixture deficient in the dopamine precursors tyrosine and phenylalanine," subjects show suboptimal performance in the Wisconsin Card Sorting Task (which requires attentional shifts across visual dimensions).

Returning to the neural circuit for eye movement control, we must make space for attention, as another name for a sensitivity mechanism that works from the top down and is oriented to more or less explicitly rewarded goals. It would involve quite a bit of interaction among cortical areas, with dopamine input modulating prefrontal cortex, which, in turn, exerts control over spatial coding in the parietal cortex and multiplicative scaling of sensory features in posterior cortical areas (Buschman & Miller, 2007; Rossi et al., 2007; Tomita et al., 1999). This top–down pathway would then continue from parietal cortex onto early sensory areas (Saalmann, Pigarev, & Vidyasagar, 2007), so that we are left with a complicated, hierarchical system consisting of cascaded projections that feed sensory information forward and abstract, contextual, task-relevant information downstream. In figure 2.6, all of this cortical interaction would squeeze into the excitatory projection to superior colliculus. Where I put the frontal eye field as a single node that receives dopamine input and drives the collicular response, this is really just shorthand for the intricate cross talk among cortical areas. For cross talk there is, certainly, for sure, and undeniably.

Sometimes it seems the whole brain is busy doing something in relation to reward, always chattering about the sugars of life, though not in the mass kind of functional way envisaged by Karl Lashley. If the neural structures and their neurons all play their little idiosyncratic part in the same piece, it should not surprise us, for the piece is a big one after all. It is our Symphony Number One.

The Pursuit of Happiness

One of the three unalienable rights mentioned in the U.S. Declaration of Independence, the pursuit of happiness belongs to the domain of what is self-evident, must be true, and cannot be questioned. Yet, question we do despite our best intentions to be happy with the truth. It must be that the old truth is never complete or detailed enough when we wish to know more about how it actually works, and whether it takes a Y, as it did in the 2006 movie with Will Smith, or whether it really refuses to take a "Why?" question. With Alistair MacIntyre in *A Short History of Ethics* (2004, p. 60):

In a choice between goods, if happiness were offered along with one but not the others, this would always and necessarily tilt the scales of choice. Thus, to justify some action by saying "Happiness is brought by this" or "Happiness consists in doing this" is always to give a reason for acting which terminates argument. No further *why?* can be raised. To have elucidated these logical properties of the concept of happiness is not, of course, to have said anything about what happiness consists in.

That happiness is desirable would be a self-sufficient matter of Aristotelian logic, an agreement built into the construction of our semantic system, a lemma given in the language that we share. Anyone not choosing the happy option would be making a mistake, due to lack of knowledge about the world, faulty perception, or inefficient motor control during the execution of the approach behavior. Or there might be something else going on, in the pursuit of an ulterior form of happiness, when the endurance of hardship is aimed at a better fate for the self in the paradise, heaven, or next cycle of life as promised by a religious system. Sometimes the objective of a better life does not concern the self but the next of kin, the offspring, or a group member, however little the resemblance among the gene profiles of the one who suffers and the one who benefits. Finally, we can think of a wide variety of deviations from the prescriptive or normative paths to happiness, when unusual behavioral patterns bring more or less defensible, or sometimes downright indefensible, forms of pleasure. The perversions and destructive hedonics may be explicable as computational processes gone awry (see Redish, 2004, for a formal model of how dopamine signals fail to adapt in the case of drug addiction). However, even for the most malicious sadist or hopelessly self-mutilating wretch, there will be some kind of preference that tips the balance in favor of one thing over another.

In all of these cases, from the normal to the bizarre, happiness appears as the final tendency of choice, the systematic goal of informed decision making. As a pleasure principle, it is older and more basic than its Freudian nephew, encompassing Eros as well as Thanatos, the conscious as well as the unconscious. As for the question that cannot be asked, creationists use the agency of capitalized entities (Allah, Chaos, God, Jehovah, Tao, the Trimurti of Brahma, Vishnu, and Shiva—there are many options, though usually only one creation myth per creationist) to explain the birth of the happiness axiom in ways that are largely incompatible with any rational discourse that requires theories to be testable and wishes data to be collected.

For the inquisitive among us, the question is really how to avoid an infinite regress of pleasure principles, as we need a pleasure principle of the pleasure principle to explain the pleasure principle and so on. Life is surely better with a pile of pleasure principles than without, but until Darwin came along, the inscrutable etymology of things gave the mystics, shamans, and priests free reign to cook up any genesis. Then came natural selection, "a stable system of explanation that does not go round and round in circles or spiral off in an infinite regress of mysteries" (Dennett, 1996, p. 25). Applied to the pleasure principle and the logic of happiness, we can begin at random with the most horrific chaos and totally arbitrary information processing. Many generations later we will find that some organisms fare better than others, and that the different luck in life is not merely due to luck at all. There is method in madness, and logic in happiness.

Thus, we run into the topic of inclusive fitness, again, and how Skinner's idea of reinforcement took the Darwinian perspective to the level of behavior. However, evolution did not stop with the kinds of physical inputs and outputs that were the focus of behaviorism. The idea of happiness as a goal worth pursuing can take hold of the mind and generate wishes, prospective biases, and goal-oriented sensitivity in complex and abstract ways, sometimes quite far removed from the real world and its primary rewards, as when Residents get absorbed by their fortunes in Second Life (an Internet-based virtual world video game, developed by Linden Research Inc.). Here, I think Ludwig Wittgenstein (2003, p. 128^e) had it fundamentally wrong in one of his *Philosophical Investigations* when he stated the following:

> 574. A proposition, and hence in another sense a thought, can be the "expression" of belief, hope, expectation, etc. But believing is not thinking. (A grammatical remark.) The concepts of believing, expecting, hoping are less distantly related to one another than they are to the concept of thinking.

Believing is thinking, and the concepts of believing, expecting, hoping are as closely related to one another as to the concept of thinking. With happiness as the final tendency, the various forms of cognitive processing compute probabilities and possibilities in an effort to discern what is good and how it can be reached, acquired, or made real. For wishes to come true, the concept of thinking should be allowed to play an active role, not simply expressing beliefs and hopes but actually being one of a set of mental activities that shape each other, so that in effect beliefs and hopes become the expressions of thought, which can, in turn, be expressed in propositions, which invite a new round of believing, wishing, thinking, and so on.

In the quoted paragraph, Wittgenstein put an obscure grammatical remark between parentheses, without filling it in or making it explicit. My best guess is that he was thinking of an overarching grammatical structure, in which the beliefs or expectations would be embedded in the clause that follows after "I think," as in "I think I would like to see it happen that…" To me, it rather looks like Wittgenstein missed the point about the grammar of thinking, as an activity performed by a subject, an individual, or an ego—a grammar that is no different from that of hoping, expecting, or wishing. There is no such thing as free thought, unbounded by a subject or a subject's being in the world. In this sense, the error of Wittgenstein may be akin to that of Descartes, as argued by Damasio (2005). More than merely a blind spot for the body and the necessary grounding of an individual's experience in the here and now of biological, historical, and cultural contingencies, Wittgenstein's separation of thought from other mental activities is a deliberate attempt at casting the cogito as a formal operation that would be rational, neutral, and innocent of semantics. While this may open the door to artificial intelligence and the invention of a fully logical mind–machine, it fails to deal with things like purpose, goal, and happiness.

Perhaps Wittgenstein's position is not even all that unusual as it opposes the purity of what is rational to the irrational mess of beliefs, wishes, fairy tales, nonsense, chaos, and whatever else fails to be absolutely logical and deterministic. Where does meaning fit in this scheme? If rationality is to be a fully reliable set of formal operations, it should be uncorrupted by meaning. Yet, meaning cannot very well belong to the family of the irrational either; that would be absurd—a kind of absurdity that might generate radical philosophies enmeshed with pessimism and nihilism. There seems to be no easy way out between the pure but empty form of reason and the unpredictable noise of all the rest.

Instead, with thought intrinsically bound to the subject, structurally condemned to a personal perspective, and inevitably biased by limited experience, we can bring together all levels of thinking and wishing in a camp dedicated to happiness as the final valuation and the ultimate meaning, squarely opposed to the absolute nothingness of chaos and randomness. Doing so, we will find different degrees of meaning and truth in any kind of statistical regularity or systematic variability through the mental activities of thought, belief, expectation, hope, and so on. Rational thinking in this scheme would represent more extensive information processing, comparing one belief against another, pitching one emotion against another. It contributes to decision making by employing complex forms of computation and more elaborate ways of weighing the options, but it is essentially governed by the same orientation—yes, *bias*—toward happiness as all other forms of mental or behavioral activity. If anything, rational thought promises to be more efficient at making good decisions and bringing forth happiness and so could be seen as the champion emotion, the mother of all emotions.

The anatomical data agree. The prefrontal cortex, being the undisputed home of the most complex executive and cognitive functions (Roberts, Robbins, & Weiskrantz, 1998), receives a massive and crucial dopaminergic input that, if nothing else, presents an excellent opportunity for a fundamental orientation to reward in the highest of the highly intellectual transactions of the mind. The best evidence for a neural mechanism of logical inference comes from dorsolateral prefrontal neurons in a paradigm in which monkeys learned to propagate reward information to new objects in a category (Pan et al., 2008). Orbitofrontal cortex, on the other hand, is consistently called into action to develop a more sophisticated or elaborate understanding of the rewards available in a given situation, be it in monkeys, when they compare a possible outcome with other alternatives in a given context (Tremblay & Schultz, 1999), or be it in rats, when they learn to associate compound stimuli with specific outcomes (e.g., receiving banana- or grape-flavored sucrose pellets) rather than generally categorizing the outcome as "good" or "positive" (Burke et al., 2008).

Also, in human fMRI studies, the orbitofrontal cortex emerges as a brain structure that computes reward information relative to a context and develops some kind of abstract currency or metric to rank different outcomes in order of preference (Elliott, Agnew, & Deakin, 2008; O'Doherty et al., 2001). Fruit juice is the default primary reward, and money the prototype abstract reward, but more social incentives such as hierarchical status (Zink et al., 2008) or reputation (Izuma, Saito, & Sadato, 2008) also get the complex reward circuitry going, with subtle differences in contributions from different prefrontal structures as well as the dorsal striatum.

With factors like social reputation, we find ourselves firmly in the field of neuroeconomics, as humans go to great lengths to secure positive comments by their peers for any transaction or service performed. A quick look at how eBay works drives the point home. (An introduction is probably not needed, but to be sure, eBay is a rather notorious multibillion-dollar operation that provides a Web site for online auction and shopping.) No one wishing to be relieved of a product can afford a blemished record under *meet the seller*, and traders may occasionally be bullied into excessive efforts at being nice and friendly when a buyer shows the slightest sign of dissatisfaction—the customer more a tyrant than a king. An fMRI study by Delgado, Frank, and Phelps (2005) showed how the reputation of hypothetical trading partners affects the actual choices as well as the concurrent BOLD responses in the caudate nucleus of subjects who participate in a "trust game," one of the classic paradigms in behavioral game theory (see Camerer, 2003, for a comprehensive introduction).

In the trust game, subjects are asked to make a first move, either keeping all of an allotted amount of money or sending it to their partner. If the subject chooses to send the money, the bank or the house multiplies the sum, leaving a sizeable treasure in the hands of the partner. The partner can then choose to be selfish and smoke a big cigar or be a good sport and return a decent portion (more than the original amount invested) to the subject. Not surprisingly, subjects, when moving first, are more likely to choose the risky option and invest the money with partners who have excellent credentials than with known cheats. The trust game quickly became a favorite among researchers interested in the neural mechanisms of cooperation and reciprocity (King-Casas et al., 2005; McCabe et al., 2001), and Paul Glimcher (2003) highlighted the immense potential of the paradigm in his book that defined neuroeconomics.

Under the rubric of neuroeconomics, I will gladly chime in with the zeitgeist; there is a wealth of interesting research on reward-oriented processing being conducted. However, it is also somewhat astounding, a bit baffling, and perhaps slightly disconcerting to see how the field suddenly took off in the space of a few years, seemingly urging many researchers to drop what they were doing and join the hype. The beginnings were relatively slow, with a handful of Wurtz-like studies that checked how reward probabilities matched with the choices made by monkeys in eye movement tasks (Coe et al., 2002; Platt & Glimcher, 1999; Sugrue, Corrado, & Newsome, 2004). This work revived the notion of the Matching Law, as introduced several decades earlier by Richard J. Herrnstein (1961), a good Skinnerian, probably best known for his controversial book with Charles Murray, *The Bell Curve* (1996). The basic idea of the Matching Law was simple enough. The likelihood of choosing an option among a set of alternatives would match the likelihood of reward for that option, relative to the reward rates for the other alternatives. The Matching Law can be seen as a prescriptive formula for rational choice, allowing us to examine how and when choices deviate from the standard (e.g., Baum, 1974). For behaviorists, this is old news indeed. However, when neurophysiologists started exploring correlates of the Matching Law in the brain, it piqued the interest of the exponentially growing fMRI community. At first, researchers simply tracked hemodynamic responses to random rewards as subjects performed gambling tasks (Delgado et al., 2000; Elliott, Friston, & Dolan, 2000), but the Matching Law was next on the agenda (e.g., Haruno et al., 2004).

Soon it dawned on economists, social psychologists, and neuroscientists alike that neuroeconomics and fMRI was a place where everyone could meet. More sophisticated measures are being devised to distinguish, for instance, goal values from reward prediction errors (Hare et al., 2008) or risk with known probabilities from uncertainty with unknown probabilities (Huettel et al., 2006). Concepts such as altruism are now investigated in the scanner, my favorite study being the one that reports warm-glow motives in charitable donations (Harbaugh, Mayr, & Burghardt, 2007). And most promisingly, the groundbreaking work by Daniel Kahneman, Amos Tversky, and their associates (Kahneman & Tversky, 2000; Kahneman, Slovic, & Tversky, 1982) is beginning to be translated to fMRI research (there is a study by De Martino et al., 2006, which I will come back to in chapter 3).

All these are wonderful developments, of course, but I cannot help occasionally feeling a little exasperated at how vague and general the computed brain correlates can be in these studies. Surely the sentiment reveals a defect in my own person—as an undergraduate student at the University of Leuven in Belgium I did rather well in most courses, if I may say so myself, but managed to flunk exactly one, *Social Psychology*, and barely scrape by in another, *Economics*. I guess I would prefer to see more detailed comparisons between behavioral variability and neural codes, of the kind shown in an fMRI study by Pessiglione and colleagues (2007) on how the brain translates money into physical effort. And, of course, it is early days still in fMRI research. The methods are bound to be optimized, both in the resolution of the magnetic resonance signal, the analysis of the hemodynamic responses, and the sophistication (or perhaps preferably *simplification*) of the behavioral paradigms.

At present, it appears as though neuroeconomics has lost touch with the rest of neuroscience. I certainly would be happy to see an increase in fMRI studies that aim to get relatively close to behavioral paradigms that we can combine with physiological, pharmacological, or anatomical techniques that rely on animal models—relatively close, for the benefit of convergence and continuation.

But perhaps not too close. I can think of at least one study with fruit flies (Andretic, Van Swinderen, & Greenspan, 2005) of which we probably do not need a human fMRI version, considering the invasive techniques employed to create sexual bait. "A single male was presented with a virgin female (either intact or decapitated, see Experimental Procedures)," the authors explained (p. 1168). Fruit flies on speed (methamphetamine, a drug that stimulates dopamine receptors) started "courtship" toward decapitated females faster than did sober fruit flies—a clear effect of sexual arousal, or did the dopamine stimulation make romantic wishes seem to come true? In any case, it appears that the visual system never got the chance to inform the rest of the fruit fly brain that it was about to descend into necrophilia—from wish come true to fear materialized.

Maybe the anticipations do in fact easily waver between positive and negative, like in the following wonderful poem by the true British master of innovative writing, J. H. Prynne (2005, p. 335). The poem lives in *The Oval Window*, a sequence especially written for neurophysiologists, I like to think (as all neurophysiologists but few poets know, the title does not refer to any office, but to a little hole in the skull, covered by a membrane that separates the middle ear from the inner ear—not the eardrum, but deeper inside). I will not try to paraphrase or interpret the poem, and I hope that the words may speak for themselves about the pleasures and pains of prospective processing. Let it be a proper bridge between chapter 2 and chapter 3 (please read slowly, in small portions, and feel free to indulge in a second or a third time):

Drawn to the window and beyond it,
by the heartfelt screen of a machine
tenderly lit sideways, the wish to enter
the sea itself leaves snow dark as sand.
Pear blossoms drift through this garden,
across the watcher's vantage clouded
by smoke from inside the hut. Tunnel
vision as she watches for his return,
face and flower shining each upon the other.
   So these did turn, return, advance,
     drawn back by doubt, put on by love.
Sort and merge, there is burning along
this frame; and now before you see
you must, we need its name.

3   Fear Materialized

> But his doom
> Reserv'd him to more wrath; for now the thought
> Both of lost happiness and lasting pain
> Torments him; round he throws his baleful eyes
> That witness'd huge affliction and dismay
> Mixt with obdúrate pride and steadfast hate:
> At once as far as Angel's ken he views
> The dismal Situation waste and wild.
> A Dungeon horrible, on all sides round
> As one great Furnace flam'd, yet from those flames
> No light, but rather darkness visible
> Serv'd only to discover sights of woe,
> Regions of sorrow, doleful shades,...

The fall from Heaven, in these paradoxically ecstatic lines (Milton, 1989, p. 6), sounds cruel in the most magnificent way. John Milton, once the proud Secretary of Foreign Tongues, fluent in dead or buried languages such as Latin and Dutch, dictated the verses of *Paradise Lost* in utter misery, a middle-aged man completely blind and persecuted by the powers that be, in the early years of the Restoration, when Charles II became the King of England after a decade of republican government. *Paradise Lost* first appeared in print in 1667, only a year after the Great Plague, which killed roughly one in five Londoners. They were hard times indeed. Yet, somehow the undeniable confidence and eloquence with which the horrible words proceed ("with obdúrate pride and steadfast hate") call for another interpretation, one in which pain and vulnerability make room for urgency and determination. Fear, materialized, refuses to be a reason to give up hope. Even if the faculty of vision conjures up nothing but images that belong to the worst case scenario—the whole "dismal situation waste and wild," everything just "one great Furnace flam'd"—this is no sufficient cause for submission.

The main character portrayed in the quotation, the angel newly fallen, better known by the name of Satan, is only at the beginning of a further set of adventures, which will include more of the same, with another creature ("Man") falling from a sweet place ("Paradise"), a perfect theatrical example of the Eternal Return before anyone put a name to it (be it Deleuze, 1994, or Friedrich Nietzsche as transformed in Deleuze, 1994). We can only speculate about whether Milton recognized a bit of his own predicament in Satan's. Neither the fallen angel nor the former civil servant had a happy ending waiting at the close of the story of life, but both were tireless warriors, who found the energy to create a masterpiece from the depths of despair. For Satan, it was the design of the original sin; for Milton, the sublime poem that painted the occasion in words.

William Blake, the oldest and most eccentric of the six great Romantic Poets, famously declared a good hundred years later that "[t]he reason Milton wrote in fetters when he wrote of Angels & God, and at liberty when of Devils & Hell, is because he was a true Poet and of the Devil's party without knowing it" (1975, p. xvii). The statement stands as a glowing reference, an unmistakable compliment, even if it is addressed only to Milton's unconscious mind. Elsewhere in *The Marriage of Heaven and Hell*, Blake corrects Descartes's error, gives the body its due, and notes the necessity of contraries in human existence—"Attraction and Repulsion, Reason and Energy, Love and Hate," all is part of life, with Good defined as "the passive that obeys Reason" and Evil, "the active springing from Energy" (p. xvi). If hate, repulsion, and pain go together in the domain of hell, there may be something to be said for a solid dose of energy to turn things around. Ultimately, as Blake fully appreciated, the active springing must be a good thing. Fear, materialized, works for the benefit of the same body that wishes to see other things come true.

Round He Throws His Baleful Eyes

Whether foreboding evil, and intent on harm, or more a victim than a culprit, and consumed by sorrow, Satan is busy vigorously making saccadic eye movements. The active springing, set in motion by the advent of disaster, has an arousing effect that pricks up the Devil's ears (in the usual, pre-Joe Orton sense). Curiously, the behavioral and physiological expression of fear seems to lower the threshold for things to impinge upon the senses, as if fear begets more fear, when the frightened mind believes to be witnessing further nasty materializations among the onrushing stream of hardly processed sounds and images.

A fascinating study by Susskind et al. (2008) speaks directly to the issue. They started by carefully observing that the facial patterns expressing fear versus disgust exhibited markedly different tendencies in terms of "skin surface deformations" (p. 844). Fear raised the eyebrows, increased the eye aperture, and elongated the nose, whereas disgust did the exact opposite. No doubt many wise people had noticed something of the sort before, but Susskind et al. suddenly made it make total sense with the most elegant and simple hypothesis. Fear, they suggested, looked like a behavioral response that serves to boost further sensory exposure. Disgust, on the other hand, would allow the affected party to dampen any further processing of the nauseating object that evoked the emotional response.

Several measures supported the proposal of antagonism between fear and disgust. When subjects were asked to mimic fearful expressions, they had the impression of being able to see more of the visual field in front of them than when they were trying on an expression of disgust. With fearful expressions, the subjects also showed an increase in the speed of eye movements and breathing. Taken together, the data made a clear case that there was something systematic about the way in which the emotional expression played with the senses. Happy though I was to read the delightful empirical report, I do wish to add a note of caution with respect to the title and the general interpretation of how the emotional expression affects the sensory processing.

Susskind et al. (2008) have it in bold letters (as their title) that "Expressing Fear Enhances Sensory Acquisition"—a slogan that sounds suspiciously like advertising the discovery of a sensitivity mechanism, implying an improved signal-to-noise ratio. But which signal would the improvement pertain to, and over what kind of noise would it prevail? I would rather think that the expression of fear biases the perceptual system toward information processing in general. Applying the LATER vocabulary, I think the effect would be to raise the starting point for any signal, bringing every possible input closer to the threshold of perception. Simply by having the doors open, the nostrils wide and the eyes practically popping out of the skull, the terrified soul would be more likely than her or his sickened sibling to take note of whatever presents itself for processing. If anything, I would predict a more specific fear-related bias mechanism in parallel with the opening of the gates, with selective anticipatory processing in neural structures that represent the most dreadful objects.

In all other respects, the account of different facial expressions and their functional implications deserves the status of instant classic among research papers. The authors rightfully claim that their "convergent results provide support for the Darwinian hypothesis that facial expressions are not arbitrary configurations for social communication" (Susskind et al., 2008, p. 843). The natural philosopher developed the hypothesis in *The Expression of the Emotions in Man and Animals* (Darwin, 1965), first published in 1872, still a highly absorbing and instructive read with plenty of painterly rendered anecdotes that illustrate the arguments very effectively. One of the central tenets is that the emotional expressions were shaped from more complete behavioral responses, aimed at direct or physical interaction with an emotive stimulus, such as spitting out a rotting apple, hooking your teeth in the fur of an all too cocky competitor, or acquiring a little pat on the back from your master, the three hundred pound gorilla you do not wish to upset.

Less energy would be expended and less damage done if the behavioral responses do not have to be executed in full. Perhaps showing your teeth, merely signaling your readiness to bite, might be enough to convince the all too cocky competitor that a little detour is actually a good idea—a clear win–win solution, in which you do not have to work up a sweat and the useless brat has no wounds to lick, no pride hurt. Formally, the expression would be no more than a relic, a trace, an abbreviated version of the behavioral response. Functionally, the same end is reached with cheaper means. However, between form and function, there must have emerged a third party, linking the one with the other. The behavioral relic has become an expression, a sign or a signal that has to be perceived and interpreted to achieve the desired physical effect. It is probably fair to call this a form of communication, either from one individual to another or via a reflexive arc back to the self.

Is this a relevant evolutionary theme for the ontogeny of language? The emotional expressions certainly do convey information in a format that, in principle at least, sets up a relatively easy path to abstraction and deictic relations. The case in point was provided a while ago, when Seyfarth and colleagues (1980) reported, from field observations in Amboseli National Park, Kenya, how vervet monkeys give different alarm calls that distinguish between leopards, eagles, and pythons. The monkeys also respond with relevant actions to the different calls, as observed when the researchers played back the recordings with no predator in sight (presumably the researchers themselves were properly camouflaged). For a leopard call, the monkeys run up a tree—something they would be wise not do for an eagle. This is already one step beyond what Darwin talked about.

However, even in the more basic emotional expressions, the beginnings of symbolic representation are in place with behavioral relics that tend to gradually lose their resemblance to any original action pattern—to the point that it now requires a *hypothesis* to claim that the expression started out as something else than an arbitrary configuration for the purpose of social communication.

In any case, the gist of the story is clear: From vervet monkeys that do or do not run up trees to subjects that raise their eyebrows, increase the aperture of their eyes, and elongate their nose, the expression of fear seems to have a purpose, investing energy with future benefit in store, if not explicitly in mind. And well it might have a purpose, for why would we bother with fear if there is no hope for delivery from pain? Darwin (1965, p. 81), ever the ideal observer, noted more neutrally:

Pain, if severe, soon induces extreme depression or prostration; but it is at first a stimulant and excites to action, as we see when we whip a horse, and as is shown by the horrid tortures inflicted in foreign lands on exhausted dray-bullocks, to rouse them to renewed exertion. Fear again is the most depressing of all the emotions; and it soon induces utter, helpless prostration, as if in consequence of, or in association with, the most violent and prolonged attempts to escape from the danger, though no such attempts have actually been made. Nevertheless, even extreme fear often acts at first as a powerful stimulant. A man or animal driven through terror to desperation, is endowed with wonderful strength, and is notoriously dangerous in the highest degree.

Fear and its sensory cousin, pain, produce an inverted U shape in the behavioral response, from most powerful at moderate levels of fear and pain to capitulation and passivity at extreme levels. Total catastrophe, after the fact, elicits unconditional surrender, leaving only one option for happiness, that is, to somehow internally transform the situation—accepting, reevaluating, denying, or in any other way modulating the meaning of the disaster. The defense happens in the head, by means of cognitive mechanisms aimed at rewriting history, or taking a contemplative stance, saying things are really not that important or could have been worse. The trick, then, is to shift the focus, think of something else, or look on the bright side, the way Brian managed to do in such an unlikely way that the effect is positively hilarious, if somewhat sacrilegious, in Monty Python's *The Life of Brian*, the 1979 comedy film that traces the many misadventures of a Messiah look-alike until he ends up being crucified, singing an infectiously cheerful tune, suggesting that we should "always look on the bright side of life."

However, it will be better yet if we do not have to resort to revisionist logic in the pursuit of happiness. In the region of the evitable, when doom is only prospective and uncertain, action is warranted, and the body duly manages to muster up whatever resources it has within itself to curb the danger or prevent further damage. The emotion goes beyond its expression and achieves more than being just a signal of potential adversity. It activates the appropriate physiological mechanisms to engage in the most powerful response, but only in those circumstances in which there is room for the response to be successful.

Presumably, natural selection and the metaphorical teleology of inclusive fitness arrange for good organisms to know intuitively which fights to pick and when to quit. In the end, the strength of the emotion biases the behavioral strategy.

There is an unmistakable economy that governs the behavioral ramifications of fear and pain, with costs and benefits analyzable in terms of invested energy and differential outcomes, to be ranked in order of proximity to happiness. Surely this economy is one of the basic truths of the human condition, forcing us forever to juggle upshots and downsides, nothing being perfect, no such thing as a free lunch, or as Jane Fonda used to put it, refreshingly bluntly, in her aerobics videos in the early 1980s, "no pain, no gain." The positive and the negative, the desires and fears, eventually must translate to the same currency if decision making is to incorporate the entire complexity, or even only the most salient handful of aspects, of any given situation. To weigh different options efficiently, we have to integrate all relevant features and implications, some good, some bad, and so we need a single metric that can deal with any departure from the zero point of neutrality. Eventually, the mechanisms that make wishes seem to come true, and let fears appear to materialize, must converge onto shared neural circuits. A fruitful approach to the study of decision making, then, is to try to map the degree of overlap.

The Singularity of Restlessness

For the Portuguese poet Fernando Pessoa, creator of dozens of heteronyms, complete with distinct personalities and detailed life histories, there was one, Bernardo Soares, who integrated all energies in Pessoa's masterpiece, *The Book of Disquiet* (2002). Perhaps Soares was the heteronym closest to the real Fernando Pessoa, closer even than the heteronym Fernando Pessoa (it does get a bit complicated when the poet uses his own name as a heteronym). In the singular voice of Soares, *The Book of Disquiet* became at once the most comprehensive and the most private of all of Pessoa's works, a chaotic collection of paragraphs, consisting of musings, dreams, aphorisms, frenzied revelations, dry criticisms, and frank observations that derive from merciless introspection, never entirely happy or sad with anything but always affected, moved, and ready to read meaning, and a bit of humor, into the banalities of everyday life.

Without wishing to write a journal or communicate a history, Pessoa/Soares noted: "I, a pathetic and anonymous office clerk, write words as if they were the soul's salvation" (2002, p. 15). The statement might sound a bit megalomaniac at first, especially if we gloss over the mere likeness of "as if" or the suggestion that salvation and soul are simply figures of speech. Or perhaps we are right to gloss over the likeness of "as if" and should consider the statement not presumptuous, but properly heavy, coming from one of the literary giants of the twentieth century, who lived in tragic isolation and virtual obscurity throughout his entire life. His literary endeavors took place below the radar, underground, and were not aimed at external recognition or rewards. The quest for salvation was, at least possibly, genuine, even if the poet thought of it as a figure of speech. The restlessness was undeniably there, driving the poet to write and to create a literary space for himself.

Perhaps all writing starts from a certain level of restlessness. Zooming out, I would even venture to speculate that restlessness is the basic state of arousal that underscores all courses of action. It represents the common denominator, or the singularity, of hope and fear, and it is principally defined by its uncertainty. In the words of David Hume (2002, p. 281): "'Tis evident, that the very same event, which by its certainty wou'd produce grief or joy, gives always rise to fear or hope, when only probable and uncertain." For Hume, it was precisely the uncertainty that set fear and hope apart among the "direct" affections. With the uncertainty came restlessness, or (still on page 281):

> Probability arises from an opposition of contrary chances or causes, by which the mind is not allow'd to fix on either side, but is incessantly tost from one to another, and at one moment is determin'd to consider an object as existent, and at another moment as the contrary. The imagination or understanding, call it which you please, fluctuates betwixt the opposite views; and tho' perhaps it may be oftner turn'd to the one side than the other, 'tis impossible for it, by reason of the opposition of causes or chances, to rest on either.

Fluctuation and restlessness are the defining features for hope and fear, so much so that the positive and negative strains of uncertainty often seem necessarily intertwined or entangled. Hope for one thing would imply fear for the nothingness of that one thing, and fear for another thing brings with it hope that the other thing in question be naught.

Sometimes it can be quite tricky to define which direction the excitement takes, whether it is hope or fear that drives our decision making. Sometimes there is a sneakily pleasurable aspect to horrific images, as the movie industry amply demonstrates, churning out plenty of new horror movies each year. No doubt a similar kind of double-edged excitement took hold of John and his fellow hikers as they walked the Kokoda Trail, in a jungle region of Papua New Guinea, on August 28, 2008. BBC News reported John's gruesome discovery of a body, probably a "World War II airman," hanging from a tree, high up in the canopy. A couple of days later the members of a new expedition, wishing to recover the corpse, found nothing more than a few moss-covered branches. It was a false alarm all right, though I cannot decide whether John had prematurely seen his wish come true or his fear materialized.

In the case of fear as well as hope, the excitement is oriented to a given object—the thing, the train of things, the event, or the sequence of events, whatever it is that you hope or fear. The object of fear, or of hope, functions as the signal in signal detection theory, the question to be answered or the hypothesis under investigation. Is it there, or is it not? Between being and nothingness, we tend to err along the side of existence—meaningfulness being our default assumption. Borrowing the voice of evolutionary psychology, we can point out that in the vast majority of circumstances the costs of false alarms are smaller than those of misses. Erroneously deciding that the desired or dreaded thing is present usually does not cost much more than a temporarily increased heartbeat, whereas failing to notice a real threat or an actual opportunity to obtain a rare reward would be a matter of certain gravity, potentially to be paid for with the highest price of life. Particularly in the fear department, the undetected can be lethal—think of a black mamba biting you or, more likely, a big truck (a V12 Detroit Diesel two stroke engine) blasting the side of your tiny car (a little Japanese K-car).

This is not to say that false alarms are by definition harmless. Gigerenzer (2002) argued very convincingly that some false alarms in medical diagnosis (e.g., erroneously being labeled HIV positive or falsely being diagnosed with breast cancer) can have a drastic negative impact on an individual's quality of life, pushing people into a depression, complete with suicidal tendencies. Even false alarms can be lethal. Nevertheless, given a choice between "There is something there" or "It is really nothing," we are usually predisposed to believe in the meaningfulness of what is before us. Perhaps the mind–brain really is a minimalist theorist, always going with the easiest good-enough explanation unless forced to admit to something more complex—an idea not unlike Herbert Simon's (1956) proposal that we "satisfice" rather than "optimize" when we make decisions, trying to find solutions that are good enough to make us happy, regardless of how close they are to the ideal or optimal. For the minimalist theorist, everything has its purpose, and every action is caused by an agency.

The bias toward meaningfulness may be due to the structure of language, with its basic formula that relates signifiers to the signified in order to move from the here and now to a virtual world that creates images for any time and any place. Perhaps we have grown so dependent on language that we have acquired its basic formula as a first intuition, or second nature, assuming everything to be a signifier for a signified and, in turn, each signified to be a signifier of yet another signified. Compulsively applying the logic of recursion, each object or event would become a reference to something else, as if the whole world is an empire of signs, and we can always ask ourselves, "What does this stand for?" "What does it mean?" or "Why is it like this?" Chaos, nothing, and coincidence are the least preferred answers.

Whoever has experienced being cornered by a four-year-old persistently asking "Why?" will know how big the temptation is to end the cycle with a final authoritative stop. The child will not take "Just so" for a satisfactory answer and is likely to resist more or less violently against "Because I say so." Invocations such as God or The Good Fairy, on the other hand, work rather suspiciously well. They call an Ultimate Agency to the scene, a perfect discouragement for further inquiry, while the main hypothesis of the minimalist theorist stays intact, and everything still has its purpose—this type of reasoning about design comes naturally to most members of our species, even children age five years old (Kelemen, 2004) or patients suffering from Alzheimer's disease (Lombrozo, Kelemen, & Zaitchik, 2007).

The preference for teleological interpretations, persistently reading meaning regardless of whether there is any in a given set of circumstances, may have implied an array of indirect evolutionary benefits, from sealing group cohesion to sustaining morale in the face of dire circumstances, operating in a fashion similar to that of placebo effects or self-fulfilling prophecies. Perhaps the power of the God meme is real, if derived from a fantasy, however visionary and grand. The point is that the prospective processing in expectations and beliefs can indeed have a measurable impact on behavior and its likelihood of success (see Beauregard, 2007, for a survey of this topic).

Looking for meaning in everything, the singularity of restlessness predisposes us to find things. "If you do not shoot, you will definitely miss" is the somewhat terse transliteration of a crisp expression in Dutch ("Niet geschoten is altijd mis"), urging the skeptic to try anyway, even if the chances of success are low. To reduce the chances of missing something good, or bad, something important whichever direction it takes, the brain is wired up to shoot for meaning at the earliest opportunity. This singularity of restlessness does not replace the concepts of arousal and valence but rather emphasizes that there are important commonalities among the mechanisms that underscore the hopes for reward versus the fears of calamity.

Arousal, as a measure of the intensity or amplitude of an emotional response, and valence, as an indicator of the positive or negative direction of the emotional response, can be combined to form a bipolar continuum from most happy to most sad (Russell, 1991), though more recently several researchers have proposed a model in which emotions are coded on a V-shaped axis with neutrality at the bottom of the V, and the highest positive and negative valences producing the twin peaks (Cunningham, Raye, & Johnson, 2004b; Winston et al., 2005). Data from fMRI investigations, based on primary sensory processing of taste (Small et al., 2003) or smell (Anderson et al., 2003), as well as more abstract, cognitive happenings while reading single words (Lewis et al., 2007), have consistently shown dissociations of arousal and valence, with the amygdala in charge of the quantitative domain of arousal and the orbitofrontal cortex responsible for more qualitative descriptions in terms of valence.

This set of observations fits nicely with a distinction made by Roesch and Olson (2004), who mixed reward and punishment in their Wurtz-like paradigm. By the time this study was published, reward-oriented processes, including bias and sensitivity mechanisms, had already been documented in several brain areas, but Roesch and Olson correctly remarked that the reward factor in previous studies confounded the neural coding of value with coding of arousal or intensity of the motor output. To circumvent this problem, they offered their subjects (two good little monkeys at the height of their career) three different types of choice, using an eye movement task in which looking at one visual target implied being treated to an associated outcome. The choice was always between a (desirable) liquid reward and a (rather aversive) time-out, but with variable amounts of liquid and time.

To be extremely precise—this exemplary study really deserves it—the monkeys could choose between (1) a large reward and a short time-out, (2) a small reward and a short time-out, or (3) a small reward and a long time-out. The two macaques were eager enough to do the task, responding quickly and accurately, in case 1 (earning a large reward) and in case 3 (avoiding a long time-out), whereas case 2 did not seem to generate the same level of motivation. These performance data correlated very well with the activity of premotor neurons, particularly in the buildup toward the response. In contrast, neurons in orbitofrontal cortex were, for the most part, sensitive to the reward value, firing vigorously for case 1, when there was a large reward at stake, and not so much for the other cases. This activity looked very much like a synergistic mechanism, amplifying the sensory processing for stimuli associated with a high reward value.

Comparing case 2 versus case 3, the monkeys were faced with the same amount of reward, pitched against different levels of time-out. The behavioral responses clearly indicated that the weighing of options integrated the positive and negative prospects in a way consistent with the notion of a single hedonic axis. The proximity to happiness for any option somehow combines the pluses and minuses, so that a certain amount of liquid reward gets more weight if it also implies the avoidance of a long time-out. The attraction works in part by contrast, and sometimes it can go the other way, as demonstrated by Brosnan and De Waal (2003), who found that monkeys refused to work for a lesser reward than what one of their own received in return for performing the same action—arguably a precursor of the type of aversion to inequity that might be a crucial component of human cooperation (Fehr & Schmidt, 1999).

With the aversive prospect of a long time-out, the choice for the little drop of juice literally seemed more urgent. The monkeys made speeded eye movements to the target associated with reward when it was up against a long time-out, as if a crucial deadline was looming. The imminent danger elicited a kind of fervor to escape the risky situation as soon as possible, perhaps by means of a lowered threshold, or a changed starting point for the decision process. What else can a deadline do? Having brought urgency to a situation, having emphasized the importance of escape, the deadline just sits there as a cutoff at a particular value along the horizontal axis of the arrow of time. The decision had better be made before the deadline is reached—a piece of advice that translates behaviorally to the implementation of a bias to whichever action may be instrumental for avoidance. This idea has already been tried and tested and remains very much alive with empirical evidence, provided by Reddi and Carpenter (2000), fitting beautifully within the framework of the LATER model.

Still on the topic of deadline, I cannot resist mentioning an intriguing finding a few years ago, hailed by BBC News on August 12, 2004, as a gene therapy that cures monkeys of laziness. Injecting rhinal cortex with a DNA construct that decreases dopamine D2 receptor ligand binding, Barry Richmond and his colleagues noted that monkeys failed to show the typical improvement of behavioral performance toward the time of reward delivery (Liu et al., 2004). Instead, the monkeys put out their best performance throughout all trials of a session, apparently unaffected by any moment of reckoning, as if the deadline was always around the corner or reward equally close at every step along the way. Would D2-mediated processes be particularly suitable for bias mechanisms, driven by urgency or otherwise? Dopamine tends to depress the activity of target neurons through D2-like receptors, whereas through D1-like receptors dopamine interacts with other receptors, specifically those relying on the action of fast-signaling neurotransmitters (Nicola, Surmeier, & Malenka, 2000; Nieoullon, 2002; Sealfon & Olanow, 2000). The latter type sounds perfect for synergistic processing, so perhaps the bias and sensitivity mechanisms recruit different receptor families—an idea that might be worth investigating (Lauwereyns, 2008).

Thus, the narrative turns to dopamine again, not entirely surprising in a section on the singularity of restlessness, and we might as well acknowledge here that fruit flies on speed (in the very same infamous study already mentioned in chapter 2) end up partying all night long, that is, they exhibit a significant decrease in sleep amount. The title of the article put it most succinctly: "Dopaminergic Modulation of Arousal in *Drosophila*" (Andretic, Van Swinderen, & Greenspan, 2005). We should also heed the fact that in some sense dopamine had already been discussed in relation to fears in the previous chapter—fears not of the expected but of the suddenly materialized variety—under the rubric of negative reward prediction errors, as coded by the pause in dopamine activity when an expected reward is omitted (e.g., Schultz, Dayan, & Montague, 1997). The positive and the negative do appear to come together in dopamine activity, both the good and the bad working in the same currency of spikes, albeit in different directions, either an increase or decrease, above or below baseline.

It is not only in dopamine activity that we see the different poles exerting an influence. Matsumoto and Hikosaka (2007) described a set of neurons in the lateral habenula, part of a brain structure called the epithalamus, which is known to project to the substantia nigra pars compacta, where dopamine neurons live. The activity of the lateral habenula neurons looked like the exact mirror image of what dopamine neurons tend to do, with increased activity in response to stimuli that were associated with no immediate reward and decreased activity for cues that indicated a reward. Matsumoto and Hikosaka further showed that electrical stimulation in the lateral habenula produced inhibition in dopamine neurons. In all likelihood, the inhibitory projection from lateral habenula feeds dopamine neurons with their negative reward prediction error, whereas the positive reward prediction error originates elsewhere. This view is also consistent with so-called "opponent-process theories" that take the appetitive and aversive dimensions in motivation to be locked together in one and the same choreography of internal states (Seymour et al., 2005; Solomon & Corbit, 1974).

Though the finding of negative coding in the lateral habenula neurons deserves special attention, given the anatomical connection to dopamine neurons, it was not the first time that neurons were seen to be firing more for the lesser reward. Subsets of neurons in the dorsolateral prefrontal cortex (Kobayashi et al., 2002), in the caudate nucleus (Watanabe, Lauwereyns, & Hikosaka, 2003b), and in the thalamus (Minamimoto, Hori, & Kimura, 2005) responded with increased activity for trials associated with the least preferable outcome in an asymmetrical reward schedule. In some of these cases, the activity appeared quite directly to counteract reward-oriented response bias—a topic that I must return to in chapter 7, when I will be venturing into Utopia, examining the paradoxical possibility of a world without the wrong kind of bias, when bias of any kind is such a pervasive, inevitable, and ubiquitous phenomenon, a basic feature of the architecture of human thought. To briefly anticipate that discussion here, some of the negative neural signals, indicating the absence of what was hoped for, may be instrumental in recruiting resources to redress the situation, to spend more effort at turning the wheel of fortune until it stops at a better place.

So far, though, of the Wurtz-like studies mentioned in relation to fear materialized, none really inflicted anything remotely like pain on the monkey—that is, in terms of stimuli used in the operant conditioning paradigms (the invasive surgery required to prepare monkeys for these kinds of neurophysiological experiments is a different matter entirely). The negative signals related to the absence of reward, the length of the wait for an opportunity to gain a reward, or even the presence of a reward, be it a smaller one than other rewards available in the paradigm. Kobayashi and colleagues (2006), in a study under the direction of Masamichi Sakagami and advised by Okihide Hikosaka, Masataka Watanabe, and Wolfram Schultz (an all-star team of Wurtz-like scientists), aimed to include a proper escape component in their behavioral paradigm, a memory-guided eye movement task with an asymmetrical reinforcement schedule. Now the monkeys worked either to obtain a liquid reward, to avoid an air puff—that is, "a stream of compressed air (200 kPA, 100 ms duration) directed toward the face from 10 cm distance" (p. 868)—or to obtain nothing but the usual auditory tone that followed a correct eye movement and signaled the end of the current trial as well as the beginning of the next (this kind of neutral or control trial might be taken to imply an indirect reward, or the opportunity to obtain a reward in the next trial, or the trial after the next, and so on).

The neural data with this paradigm confirmed the picture that had started to emerge already before the introduction of truly aversive stimuli, or negative reinforcers in the Skinnerian sense, which serve to strengthen the behavior that prevents their occurrence. Kobayashi and colleagues (2006) catalogued several types of neurons in the dorsolateral prefrontal cortex, some responsive particularly in relation to rewards, others specifically when the visual cue implied the risk of an air puff. Both the reward-oriented and fear-driven neural codes correlated with improved behavioral performance relative to the neutral case when there was nothing much at stake for the monkey.

Soon after this study, there appeared another one that I like to think of as the perfect counterpart—not an instrumental but a classical conditioning paradigm (simply connecting a conditioned visual stimulus with the unconditioned appetitive or aversive stimulus), recording neurons in the amygdala, or the almond-shaped extension of temporal cortex, part of the limbic system, famous for its role in fear conditioning and "practically a household word" (Phelps & LeDoux, 2005, p. 175). This time, Belova and colleagues (2007) checked the activity of amygdala neurons when the monkey received one of four treatments in a factorial combination of two levels of prediction and two levels of valence, or simply a predicted or unpredicted drop of juice or air puff. Again, there were different neural families to be recognized, some neurons strongly affected by any unpredicted outcome but less engaged with predicted outcomes, other neurons more particularly focused on unexpected juice, and yet a third group of neurons partial to unexpected air puffs.

Both the studies of Kobayashi and colleagues (2006) and of Belova and colleagues (2007) support the general hypothesis of valence-dependent coding, with both positive and negative codes in the same brain area. Yet, they also show something that cannot (or cannot yet, given the current state of the art) be read out of BOLD responses in fMRI studies. There is a certain idiosyncrasy to the firing patterns exhibited by the neurons. Some kind of taxonomy can be drawn up and should be characterized further with respect to anatomy and neurochemistry. The fact that there do appear to be subsets of neurons with distinguishable activity profiles raises the possibility that there are distinct neural circuits even within the same brain area. The data are certainly compatible with the idea that there are separable mechanisms for hope versus fear, served by neural circuits that, even if the wires look to be totally chaotic and entirely entangled, are actually fed by different anatomical inputs and are feeding into, again, different sets of clients.

The singularity of restlessness, or the common ground between mechanisms of fear and hope, is perhaps best understood as a matter of similarity in form. Though the data are still scarce and incomplete, my best guess is that the brain harbors fear-dependent bias and sensitivity mechanisms that essentially perform the same functions as the reward-oriented ones introduced in chapter 2. Some of the same brain structures and neurotransmitters will be involved, notably prefrontal cortex and the basal ganglia, glutamate and GABA, and other players may have a more dominant role for fear, notably the amygdala and the periaqueductal gray, acetylcholine and serotonin. I expect we will get to learn more in the next few years about anticipatory mechanisms that produce additive scaling effects in fear-related information processing, building on from preliminary studies that already give a hint in that direction (e.g., the fMRI study with a virtual predator, looking suspiciously like Pac-Man; Mobbs et al., 2007). Similarly, we are at the beginning of the learning curve with respect to sensitivity mechanisms that imply synergistic processing and multiplicative scaling, with improved signal-to-noise ratios for aversive stimuli.

That such sensitivity mechanisms do exist was demonstrated most clearly in an fMRI study by Li and colleagues (2008), showing how the human brain learns to recognize new smells if they predict the presence or absence of an electric shock, whereas our organ of thought remains indifferent to other new smells that have no special meaning, make no interesting predictions, and are, in molecular terms, equally difficult or easy to recognize. Or is this a special case, a different kind of sensitivity mechanism, one that involves not merely an ad hoc interaction between fear and sensory input but something more fundamental, something more structural? It is about time, I suppose, that we consider the issue of neural dynamics and plasticity a bit more closely. How soon can the signal-to-noise ratio adapt to a new context; how easily is it regulated up or down? How quickly do the neural circuits change the weights of this or that option during the weighing of options?

Snake Found Snuggled in Armchair

Some fears have so archetypal an aura to them that they seem to have been with us from the very beginning or as far back as we can remember. I can easily conjure up a vivid image, a dream image, set in the backyard of the house of my childhood in Antwerp, Belgium. It was a relatively small backyard, just a bit of slightly unkempt lawn with a large shed at the back, the aquamarine paint flaking off, a rather sad thing, which my parents did well to tear down at one point, but in my dream it was still there, at the other side of the lawn littered with coiled-up snakes, literally dozens of them, mostly pythons and boas, I think, but none very large and all awkwardly motionless, asleep, or merely waiting. Of course, there were never any snakes there in real life, not in the quietly boring and reassuring place where I grew up. I must have borrowed their visual features from a visit to the Snake House in the rather ominous Antwerp Zoo (for the exact mood, read the positively mesmerizing novel *Austerlitz*, by W. G. Sebald, 2002), or perhaps my unconscious dream scenarist had all too eagerly studied the wonderful color plates of a book on "the wonders of nature." I think the dream has been with me since I was five or six. I dreamed it only once, but the image was so strong and terrified me so badly that it somehow kept returning remarkably easily to my mind's eye; it must be in my explicit memory, an episodic trace that almost (or indeed) has acquired the status of core semantic feature that comes naturally in tow with the list of basic characteristics for items such as "backyard" or "snake."

Not that I wish to write a novel about it or perform any Freudian dream analysis. For me, the most curious aspect of the train of images and emotions that come with the episode is how autonomous and mechanic they are, inflexible and, yet, for all their machine-like properties, somehow deeply private and personal, essentially *mine*. Nevertheless, it is probably fair to say that among human individuals there are quite many who share a similar "deeply private and personal" fear of snakes. I think I understand, and I certainly fully sympathize with, the predicament of an "almost hysterical" young woman from Portsmouth, United Kingdom, who found a 1.2-meter-long orange snake in the boiler at her house, as recounted by pest controller Lee Marshall (BBC News, May 20, 2008)—an event not as infrequent as we might wish. Lee Marshall was called to a different Portsmouth address two months later. When he arrived at the premises, he found "the snake was still sat there quite comfortable in the armchair as if it was preparing itself for an evening in front of the telly" (BBC News, July 13, 2008).

The topic of ophidiophobia, or fear of snakes, easily branches out in several directions and has already invited views and comments from widely (or even wildly) different perspectives. The combination of a rigid set of physiological and emotional responses with personal, accidental, and entirely contingent imagery goes particularly well with a blend of Ivan Pavlov's conditioned reflexes and Konrad Lorenz's innate releasing mechanisms, including especially a role for imprinting (see Matt Ridley's thoroughly enjoyable *Nature via Nurture*, 2003, for a convincing account).

Psychoanalytic accounts would explore the narrative sequence and symbolic structure, emphasizing the historical contingency of the imagery in a way that neuroscience cannot begin to try to follow but might learn from indirectly, as "the talking cure" fully capitalizes on the individuality of the other, who is thought to have a voice in charge of a unique discourse (the very nickname of "the talking cure" being famously coined by one such voice, Bertha Pappenheim, better known as Anna O. in *Studies on Hysteria*; Breuer & Freud, 2000). Something of the zealousness with which the individual circumstances are recorded in psychoanalysis could definitely benefit quite a few fMRI studies in a search for factors that produce interindividual variability, especially when the investigations turn to more complex and ill-defined cognitive processes as in recent ventures into the realm of moral judgments. A good example of a missed opportunity in this regard is the otherwise very intriguing study on infection, incest, and iniquity by Borg, Lieberman, and Kiehl (2008), which remains awkwardly oblivious to the idea, if not the fact, that people come to these topics with different mind-sets and something more complex than merely a certain degree of disgust.

More or less primed by psychoanalysis, I cannot resist adding a reference here to a perspective from contemporary (continental) philosophy on the intrinsically personal conditioning of reflexes, according to which "the whole mystical game of loss and salvation is...contained in repetition" (Deleuze, 1994, p. 6). To my ears, this sounds like a reintroduction of the Skinnerian notion of reinforcement in a somewhat baroque vocabulary, the result of reading too much Nietzsche. It gets even prettier when Deleuze notes between brackets that "forgetting as a force is an integral part of the lived experience of eternal return" (p. 8). The eternal return, here, must surely be the conditioned reflex, "conceiving the same on the basis of the different" (p. 41)—the same fear response evoked by the novel exposure to a formative example, the prototype image of a snake for the next individual. Indeed, "it is not the Whole, the Same or the prior identity in general which returns.... Only the extreme forms return" (p. 41). It is all right if your snake prototype looks different from mine, as long as we share the same structures of relationships between types of stimuli and kinds of responses, physiological and behavioral, allowing reinforced behaviors to become habits, insensitive even to contextual changes (Dickinson & Balleine, 1994), or representing automatic processes, however cognitive or acquired (Carr, 1992), just as autonomous, unstoppable, and unconscious, suddenly there and immediately gone, even before we get a chance to remember whether we had willed them into action.

How much of the architecture of relationships, associations, and connections is innate, and what exactly needs to be input ad hoc as a matter of contingency? The focus by now should fully be on questions of how nature and nurture interact—what is given in terms of neural circuits, and how experience turns on or off some processes of growth, plasticity, and change. With respect to fears and phobias, a start has already been made with the proposal of an evolved module of fear and fear learning (Öhman & Mineka, 2001), implying some degree of preparedness and hardwired preferences—biases—for the kinds of stimuli that have been fear relevant in the natural history of our species, even if present-day statistics make abundantly clear that cars should really be much scarier objects than snakes.

Snakes were certainly a very real threat to our ancestors, considering that constrictors with gapes large enough to eat mammals had evolved before a hundred million years ago and that colubroid snakes with a potent venom delivery system had appeared on the scene no later than sixty-five million years ago—a time frame that perfectly coincides with that of important changes in primate brains. The evidence was reviewed by Isbell (2006), who went on to make the radical but very well substantiated claim that the threats posed by snakes provided the crucial challenges that drove evolutionary changes in primate brains—changes previously thought to have emerged as a result of selective pressures that favored visual predation, including behaviors such as stalking and visually guided reaching and grasping. Instead, the orbital convergence (or the ability to move the eyes in opposite directions to focus on a nearby point), visual specialization, and brain expansion in general would have been all about "bewaring the beast," as Öhman (2007) put it succinctly, not to hunt it but to avoid being hunted by it.

If there should be any predisposition for fear in our species, then snakes must surely be the object of choice. Regardless of when and how exactly the wiring becomes hard in any one individual, our average human takes very little time indeed to detect a snake among mushrooms, whereas the single mushroom among a grid-pattern array of snake pictures can be quite hard to find (Öhman, Flykt, & Esteves, 2001). Not that mushrooms are all that innocent, I would think, but I will concede that the danger posed by the little fungi is only there for the taking. Öhman and his colleagues make an interesting remark or two along the way in their research report, worth highlighting in the context of a monograph on the anatomy of bias. My word-hopping eyes got stuck on how "[m]ost investigators appear to agree that there is an important distinction between preattentive and postattentive visual attention" (p. 466). Most investigators would surely do better science without concepts that hark back to concepts before they exist, for how can there be attention before attention without there being attention already? Again, I suggest replacing the vagueness of the beautiful but ill-defined concept of attention with the numerical precision of sensitivity and bias. Incidentally, the word "bias" does appear fairly frequently in the paper, to describe, for instance, how "humans in general appear to have a bias to attend to biologically relevant, threatening stimuli" (p. 475). I think the word "bias" is used correctly here, even if it is thought to operate on the needlessly vague process of "to attend to," which would be captured better by the more honestly vague process of "to process."

What Öhman and his team (2001) really wish to say in relation to attentive processing, pre- or post-, is that there is good evidence of a search asymmetry, perhaps most convincingly in their second experiment, in which the time to find a snake or a spider (a "fear-relevant object") among mushrooms or flowers ("fear-irrelevant objects") does not depend on the number of mushrooms or flowers in the display—subjects always find the fear-relevant object right away. Detecting a mushroom or a flower among fear-relevant objects, on the other hand, takes longer if there are more snakes or spiders to ignore. Data of this kind call back to mind the feature-integration theory of Treisman and colleagues (Treisman & Gelade, 1980; Treisman & Sato, 1990; Treisman & Souther, 1985), briefly discussed in chapter 1. By this theory, the evidence of parallel search, independent of the number of items in the display, suggests that the target was found without the use of any deliberate feature-integrative powers.

How is it possible to detect snakes or spiders so fast? For Öhman, Flykt, and Esteves (2001) it implies that these fear-relevant objects "include some perceptual feature that makes them easy targets," whereas "such features were not present in the pictures of flowers and mushrooms" (p. 475). The researchers are even happy to speculate that the "elementary threat features" could be "related to the typical sinusoidal snake shape and to the circular body with protruding legs of the spider" (still on p. 475). My malicious self cannot help but wonder whether this theory implies that we should be able to obtain a search asymmetry for snake-like letter S pitched against mushroom-like letter T. *Beware the sinusoidal shape!*

That is not entirely fair, I admit. Perhaps I am biased against explanations of unequally weighted information processing that go solely on the basis of sensitivity mechanisms. Here, sensitivity twice gets the benefit of the doubt, saying that special perceptual features are used to detect objects, which then, in a presumable phase II, get more "attention"—a concept that Öhman and his colleagues (2001) do not define but seem to understand in the implicitly consensual way William James (1950) did, seeing as they explicitly refer to the good man's work in the first paragraph of their paper. Attention, then, must be whatever it is that makes us see things clearly and vividly, a sure provider of improved signal-to-noise ratios.

I cannot deny the theoretical possibility that the double set of implicit assumptions is in fact completely warranted. The data give us no basis to accept or reject, or estimate the amount of systematic variability explained by, the role of sensitivity mechanisms. My point is that bias mechanisms might very well have been at play, and it would be not only premature but plainly misguided to examine the processes with blinkers on, unable to even see the possibility of lowered thresholds or heightened anticipatory processing, complete with the increased risk of false alarms. Mountain biking on my way to the campus of Tamagawa University, in a surprisingly green part of Tokyo, I know for sure I have fear-relevant bias mechanisms at work in my brain every morning. My colleagues have warned me of poisonous snakes on campus, and though I have yet to see one, there have been plenty of briefly truly scary-looking twigs and branches on the secret little back road I take, especially the last section, too narrow for cars, through dense foliage left and right, where I usually never meet another human, except occasionally the old gentleman walking his dog, who used to get an even bigger fright from me whizzing by than I do from the snake-like distractions underneath my wheels. Recently, I think the gentleman got used to me. He even waves a friendly hello. I, though, keep seeing scary twigs and branches but definitely more so on that last stretch than anywhere else on my twenty-minute mountain-bike workout (that last stretch is quite noticeably uphill, I should add, not sure if this modulates the fear in any way). The image of "the beast" is then most alive inside my skull, raising the starting point of the decision line to just a short distance shy of the snake threshold, as according to the LATER model.

Öhman and his colleagues (2001), I suspect, would be willing to concede this point, given the following admission: "[S]ome of us are familiar with the experience of the seemingly automatic focusing of attention on the snake that is resting in the grass a few steps ahead along the paths we are treading. Sometimes one may even freeze a fraction of a second later to realize that the 'snake' was merely a twig" (p. 467). The description nicely fits the by now classic proposal, courtesy of Joseph LeDoux (1994, 2002), that the amygdala, as the putative neural substrate of the fear and fear-learning module, receives sensory information through two sets of anatomical projections, one fast and crude, straight from the thalamus, and another slow and thorough, via the sensory cortices (see Phelps & LeDoux, 2005, for an update with fMRI evidence). The fast and crude processing would be the one most likely responsible for false alarms, particularly in contexts that somehow prime the possibility of a threat. If the mood of the forest is dark enough, the thalamic structures would not need anything as precise as a sinusoidal shape to get the amygdala to sound the alarm—a bit of movement or merely a blobby something might do already. And I, for my part, am willing to concede that sensitivity mechanisms are likely to be called into action following, or even to a degree in parallel with, the alarming news arising from the subcortical route.

Yet, also in the slow and thorough processing of the sensory cortices we might recognize aspects that are best understood as reflective of bias in statistical terms, some conclusions requiring less empirical evidence than others. I am thinking of the impressive series of studies by Norman M. Weinberger and colleagues focused on sensory plasticity, begun in the 1970s (first subcortically, in the medial geniculate nucleus; Weinberger, Imig, & Lippe, 1972) and continuing to this day (Leon, Poytress, & Weinberger, 2008). Weinberger wrote several important reviews, switching between snappy titles ("Retuning the Brain by Fear Conditioning"; Weinberger, 1995) and more tersely academic ones ("Specific Long-Term Memory Traces in Primary Auditory Cortex"; Weinberger, 2004). We can find our way to the details in those reviews, but here I would like to borrow some of the main ideas and run with them for a more schematic presentation with figures 3.1 and 3.2.

Let us consider the case of a neutral sensory map, that is, an array of features that belong to the same dimension and have, in the unbearable lightness of the beginning, perfectly equal weights. Let us consider more precisely a neutral map for the pitch of sounds, offered by DNA blueprint, innately, in the primary

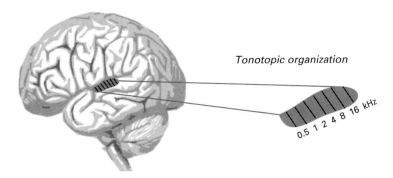

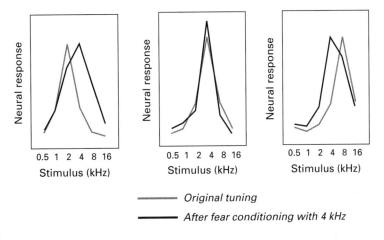

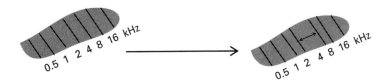

Figure 3.1
Primary auditory cortex and the effects of fear conditioning. (a) The tonotopic organization of primary auditory cortex, with a systematic anatomical layout of neurons tuned to different levels of pitch. (b) Examples of auditory cortical neurons, originally tuned to sounds of 2 kHz (left), 4 kHz (middle), or 8 kHz (right), as indicated by the gray curves. The black curves indicate the adaptation in tuning after fear conditioning with a sound of 4 kHz. (c) Tonotopic reorganization as a function of fear conditioning with a sound of 4 kHz. The area of neurons tuned to 4 kHz expands by recruiting neurons from neighboring areas.

auditory cortex of mammals, say, humans (figure 3.1a) or guinea pigs. We will be quick to note that within the neural tissue in primary auditory cortex small portions are devoted to different levels of pitch, and that this arrangement appears systematic, going from the lowest perceivable pitch to the all but lowest, and so on, climbing up the scales, all the way to the highest perceivable pitch, in a neat spatial arrangement that we call a "tonotopic organization."

Suddenly things change. Evolutionary pressures arise at the whim of an experimenter's hat such that now every sound of exactly four kilohertz is curiously followed by a nasty little bit of electric current in one of the feet of the poor subject (typically a guinea pig). The name of the game is fear conditioning, of the classical variety, with the four-kilohertz sound in the role of conditioned stimulus (CS) and the foot-shock sharing top billing as the unconditioned stimulus (US), harbinger of an unconditioned response (UR) like increased heartbeat and probably other types of UR as well, which are all bound to acquire the status of conditioned response (CR), ready to occur in what looks like a reflexive fashion following the CS, well before the senses are bombarded with the US.

Now a tone of four kilohertz can no longer pass for a neutral sound feature. Recording from auditory neurons, we would observe subtle shifts in the tuning curves (see figure 3.1b). Before the fear conditioning with four kilohertz (gray curves), the neurons would show their normal tuning, completely in accordance with the tonotopic organization—a neuron in the two-kilohertz area, for instance, will have the apex of its tuning curve for sounds of two kilohertz and will fire less and less vigorously for sounds that are further removed from the ideal of two kilohertz. After the fear conditioning, the tuning curve undergoes a slight but very meaningful shift, with the new peak at the ominous four kilohertz (black curves). This happens for neurons that from the outset were in the neighborhood of four kilohertz—neurons that previously were tuned to just one level lower (two kilohertz) or just one level higher (eight kilohertz).

Neurons further away from the dreadful four kilohertz in the tonotopic organization may not be affected all that much. The neural domain of four kilohertz simply recruited nearby tissue (see figure 3.1c), the way territorial expansion usually works, beginning with Poland and Belgium if you start from Germany. Now the fearsome kilohertz value quite literally carries more weight in primary auditory cortex. The same can be said on the basis of human brains and fMRI, as according to Morris, Buchel, and Dolan (2001), whose conditioned subjects showed stronger auditory cortical activation for fear-relevant tones than for neutral tones.

Though it must surely be too early to give any definite explanation of how the neural plasticity is actually achieved, as in which kinds of growth processes are stimulated in what kind of way, we can at least try to imagine how it *could* work. Donald Olding Hebb (2002, p. 65) suggested some sixty years ago in his classic *The Organization of Behavior* that, in modern translation, neurons that fire together, wire together. In his actual words:

When the coincidence does occur, and the active fiber, which is close to the soma of another cell, adds to a local excitation in it, I assume that the joint action tends to produce a thickening of the fiber—forming a synaptic knob—or adds to a thickening already present.

It is still the best concept around in neural plasticity. I tried my hand at applying it to the changes as a function of fear conditioning in figure 3.2. Considering that axon terminals are more stable than dendritic spines (Majewska, Newton, & Sur, 2006), I would fancy something like a neutral one-to-one mapping from the medial geniculate nucleus (or the auditory module among sensory thalamic structures) onto primary auditory cortex (see figure 3.2, top panel), which would be skewed toward a one-to-many mapping by the strengthening of existing synapses from medial geniculate nucleus neurons tuned to four kilohertz onto their cortical counterparts, as well as the loss of synapses among other thalamocortical projections and, critically, the gain of synapses between the very same medial geniculate nucleus neurons tuned to four kilohertz and cortical neurons that used to peak for a different pitch (see figure 3.2, bottom panel). To say the same without having to come up for air: Now more cortical neurons listen to thalamic news about four kilohertz.

I think this reflects a bias mechanism, but one of a more structural nature than the ones I have proposed so far. It is a mechanism that is fully in place before the arrival of sensory input, and in this sense, it agrees with the definitive feature of "anticipatory processing" in bias. Yet, in this case the preparedness is more implicit, given by the neural architecture, or the preexisting connections in the neural circuitry, without any ad hoc spiking activity ramping up toward the moment of stimulus arrival. The preparedness would still be measurable, of course, but then by more anatomical means, by sounding off the strengths and weaknesses among different synapses.

The second property, on the other hand, the additive scaling, can be examined with the neural activities of the auditory cortical neurons, comparing the gray and black curves in figure 3.1b. With respect to neurons that were originally tuned to four kilohertz, the data actually support a sensitivity mechanism. For the neighboring neurons, though, the ones originally tuned to two or eight kilohertz, we find something more radical than mere additive scaling—the data

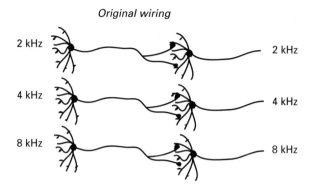

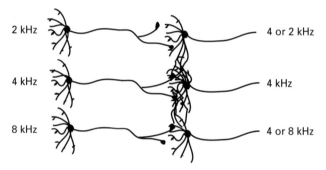

Figure 3.2
Putative mechanisms of synaptic plasticity that underlie tonotopic reorganization. The top panel shows a simplified, schematic representation of the original one-to-one mapping from auditory sensory thalamic neurons onto primary auditory cortical neurons. The bottom panel shows how dendritic growth allows auditory cortical neurons to make contact with other axonal inputs, effectively producing a one-to-many mapping as a function of fear conditioning. Now the input relating to the fear-relevant sound of 4 kHz also reaches primary auditory cortical neurons that were originally tuned specifically to 2 kHz or 8 kHz.

entail a loss of information, or a drastically altered signal-to-noise ratio, questioning even which is what. Does the erstwhile neural champion of two kilohertz now speak for two or four kilohertz? At the single neuron level, there is clearly some serious confusability. At the neural population level, when it comes to making a decision about the kilohertz value of a sound, this must mean that sounds of two or eight kilohertz are now much more likely to be mistaken for a sound of four kilohertz than before the fear conditioning. Conversely, sounds of four kilohertz now have a very easy time to win the majority vote, much easier than any other sound. In terms of the LATER model, four-kilohertz sounds are simply closer to the decision threshold than any of their competitors.

Does this account revive the notion of special perceptual features that Öhman, Flykt, and Esteves (2001) speculated about in their attempt to explain the speed of snake detection? Perhaps four-kilohertz sounds do indeed do for the amygdala of the fear-conditioned guinea pig what the sinusoidal shape naturally or instinctively does in ophidiophobia. Yet, this notion of a special perceptual feature is more accurately described as a bias mechanism—not using actual sinusoidal shapes as a way to see snakes better or more efficiently but using anything more or less similar to sinusoidal shapes to feed the bias for thinking that what you see is a snake.

Not the last word has been said about the combination of bias and fear conditioning, I bet. Just as I was going one final time through the literature that I had set aside for this chapter, there appeared an intriguing article in *Science*, connecting the two concepts in a directly political context (Oxley et al., 2008). Physiological measurements such as skin conductance and blink amplitude in response to sudden noises and threatening images ("a very large spider on the face of a frightened person, a dazed individual with a bloody face, and an open wound with maggots in it," p. 1668) covaried with subjects' political attitudes in relation to topics like gun control, capital punishment, and the Iraq war. One bias went with another—fearful people, entranced by the sinusoidal shape, voted for the policies of President George W. Bush. The authors went on to conclude, rather pessimistically, thinking the gun-toting fear was truly structural, or even determined by genetic variation, that this provided "one possible explanation for both the lack of malleability in the beliefs of individuals with strong political convictions and for the associated ubiquity of political conflict" (p. 1669).

Call me naive, but with Herry and colleagues (2008) I would think the context determines, even in mice, when to switch between exploratory and defensive behavior. Mice have distinct neural circuits for it, allowing fear to be switched on or off flexibly, if not at will. There must be hope for a similar regulation in humans (see Delgado et al., 2008a, for further cause). In any dark time, governed by the "systematic propagation of fear in the populace," complete with "[d]emagoguery, deceit, and denial of the other, all in the trusted name of 'liberty,' 'freedom' and 'democracy,' repeated mantra-like as death and mutilation reign down on untold (and unacknowledged) numbers," we will live to see a shift in paradigm if we listen to the voice of the true poet, in this case Michael Palmer (2008, with a medley of his words on pp. 30, 31, and 39), offering to "not glibly speak of a shared cultural ecology that conserves as well as innovates, that resists habits of thought and action when necessary and builds a nonnostalgic vision of pasts and possible futures, and of a less predatory present." To change the context from a program of fear conditioning to a platform for exploration and exchange, the first thing to do is to start singing a different song. Our genes will help us learn it. With more positive imagery and a reoriented bias, we may end up voting in favor of foreign aid.

Who Is Afraid of Chocolate, Plush, and Manfred?

The fear that propagates in the populace, and translates to political choice, is not, of course, specifically evoked by the imminent or imaginary danger of any actual snake. It targets not any aversive object in particular but a more general and "open" category, defined negatively, by exclusion, as everything that falls outside a trusted circle. Fear of the unknown expresses itself in various ways and to different degrees, and sometimes it relies on a peculiar hybrid of the known and the unknown, as in the case of xenophobia, or fear of the foreigner, particularly targeted at a known "unknown," that is, humans of the very well defined kind that the xenophobic individual in question does not wish to have any dealings with. Stepping back from the pathological, though, we note that fear of the unknown almost by definition correlates with wishes to stay with the known, the tried and true, "conservative" in its original sense, relating to the keeping of things as they are, or a bias toward the familiar—a topic I will explore further in chapter 4.

The bias toward the familiar produces a variety of mistakes, and may properly be called a fallacy (and so I do call it that), but we would be wise to at least consider the possibility that it has worked well for our species and even for our ancestors before they became our species. Michael Palmer (2008) subtly included the word "conserves" in his proposal for speaking not glibly—the ideal, obvious in theory perhaps but so difficult in practice, would be to balance conservation and innovation, keeping the good and introducing the better. We must have been going back and forth between the two, more or less explicitly, for at least as long as we have been trying to invent new ways of doing things, complete with fights or debates over which is the better way. Could it be that we come to the world with an evolved module to *choose*, ready to provide us with the gut feeling that we have several options? It would be a meta-instinct, operating on the basic instincts to defend what we have and to explore how we can get more, coming into play to resolve potential conflicts, to inhibit either debilitating fear or plain stupid risk taking when the context favors a temporary dominion of one instinct over the other.

Even babies less than six months of age will show evidence of opposing behavioral or psychodynamic tendencies, as when initial good-hearted curiosity turns to a violent expression of fear for the likes of chocolate, plush, and

Figure 3.3
Faced with chocolate, plush, and the hitherto unknown creature called Manfred, the overwhelmed child resorts to vigorous clamor, left cruelly ignored as parental attentions turn to recording the curious expression of fear.

Manfred, following the sudden, forced positioning on a foreign lap for the purpose of documenting a friendly visit photographically (see figure 3.3). Here, the baby's instinct of choice said, "take me back" (an action quickly performed by the photographer–parent at fault). The fear may well be prompted by an awareness of separation and could entail a component of anger toward the object of attachment. The photographer–parent had better behave for a while and might still learn a useful thing or two from the pioneer of attachment theory, John Bowlby (1990).

The fear of separation, or loss of attachment, calls to mind the more general loss aversion that represents one of the most salient and widely researched aspects of prospect theory, a highly successful set of thoughts on decision making under risk, developed by Daniel Kahneman and Amos Tversky (1979) and usually situated in the field of economics (see also Plous, 1993, for particularly reader-friendly introduction). Choices are determined by the prospects of gains and losses, with a curious but very well substantiated asymmetry such that the value function for losses is steeper than that for gains—somehow our distress with losses tends to weigh heavier than our delight about gains.

The loss aversion may get us to tend to prefer incumbent politicians over challengers (Quattrone & Tversky, 1988) or to set a ridiculous selling price for the chocolate bar in our pocket (Knetsch & Sinden, 1984). In the 1980s, researchers simply measured people's behavioral tendencies. Now, we can put subjects "in the magnet" for an fMRI study as they bid too high in an auction, so we find that the overbidding correlates with the magnitude of the BOLD response to loss in the ventral caudate nucleus (Delgado et al., 2008b).

Prospect theory, with its overt reference to the role of imagery, further predicts that preferences depend on how things are pitched, the way they are presented, talked about, and sketched, however manipulatively or unwittingly. Effects of this kind, usually discussed under the rubric of "framing," allow us, thinking optimistically, to counter essentialist views of ubiquitous fear in the populace and insist that, if only we can communicate in the right kind of frame, there is hope for prevention of irrational choices, be it overbidding for something trivial or, more damagingly, sticking with the wrong breed of politician.

Framing effects certainly do exist and can be quite dramatic. My favorite demonstration of such would have to be the fMRI study by De Martino, Kumaran, Seymour, and Dolan (2006), which had subjects choose between gambling and not gambling. The trick was in how the sure option was pitched, either as an opportunity to keep, say, £20 from an initial amount of £50 ("gain frame") or as a course of action sanctioned by the loss of £30 from the same initial amount of £50 ("loss frame"). The sure option, either framed as a gain or a loss, was offered as an alternative to a gambling option that usually had the same value in terms of utility. In the given example, the gambling option could imply a 40% chance of keeping the entire initial amount versus a 60% chance of losing all (0.4×50 plus 0.6×0, giving a total of £20 to keep, or £30 to lose, out of the initial £50).

In terms of experimental design, the crucial framing manipulation with the sure option was done extremely subtly and simply—across conditions, just two words changed, "keep £20" or "lose £30." Computationally, rationally, the choice to gamble or to go for the sure option should be immune from the antics of framing. However, behaviorally, every single one of twenty subjects succumbed to the influence of the frame, being more likely to go for the risk and gamble when the word said that money was on the verge of being lost. The prospect of something aversive, or the idea that fear might materialize, had subjects on the run, while their amygdala showed enhanced activation (De Martino et al., 2006). The data are very striking indeed, exposing the human once again as an irrational and gullible being, prone to effects of bias without a clue as to what is going on. Yet, the data are published, the researchers figured out how the bias worked—which just possibly means we could, at least theoretically, *learn* from it.

It could be that I am focused too strongly on the idea, or even the *ideal*, of learning. Disasters do happen, losses are not always averted, fears materialize for real, and no amount of books in the world will save us from sadness and depression, which is what John Bowlby (1980) had beyond the colon in the title of his third volume, *Loss*, in the trilogy that had begun with *Attachment* (the first volume, republished in 1999) and that had already morphed to threat by volume two, *Separation* (1973)—a movement in a direction opposite to the one we would wish to aspire to with Dante Alighieri (1995) in *The Divine Comedy*. No amount of books in the world (and there were many already in the seventeenth century) saved Robert Burton, man of God, bookworm, wallowing most of his life in a deep state of depression, though he did manage to write one colossal treatise on the nature of his predicament, a work published in 1621 and so brilliant and eloquent in its monomania that it is likely to remain insuperable for several centuries to come (or do we suspect I might be guilty of resorting to self-serving hyperbole, seeing as I chose the title for my own little effort here in homage of the work in question, no doubt sneakily hoping that some of the greatness, artificially inflated or not, would rub off on the present?). Here is how Burton defined the affliction in *The Anatomy of Melancholy* (2001, part I, section I, member III, subsection I, p. 170):

The *summum genus* is dotage, or "anguish of the mind," saith Aretæus; "of a principal part," Hercules de Saxonia adds, to distinguish it from cramp and palsy, and such diseases as belong to the outward sense and motions; "depraved," to distinguish it from folly and madness (which Montaltus makes *angor animi*, to separate), in which those functions are not depraved, but rather abolished; "without an ague" is added by all, to sever it from frenzy, and that melancholy which is in pestilent fever. "Fear and sorrow" make it differ from madness; "without a cause" is lastly inserted to specify it from all other ordinary passions of "fear and sorrow."

I had to look up a few words, learning as I always do with Burton: Courtesy of www.freedictionary.com, I know dotage is Middle English for "a deterioration of mental faculties" as in "senility," and ague stands for "a febrile condition in which there are alternating periods of chills, fever, and sweating" and is "[u]sed chiefly in reference to the fevers associated with malaria." But clearly the central players are fear and sorrow, *without a cause*, Burton lastly inserted, a phrase that I cannot read without a host of spurious, private, irrational associations, contingent products of learning, futile rebels, James Dean, *East of Eden*, the fall from Paradise, John Milton, writing without fetters, Satan, and the baleful eyes of the beginning of this chapter. It must be my curse, but I cannot seem to help running into bias just about everywhere. What is "fear and sorrow without a cause" other than a type of false alarm, the result of some kind of nonneutral way of weighing things, the likelihood of objects and events, the costs and benefits of options and courses of action? Not that I am reading depression in any new way—in fact, it is decidedly mainstream, more than twenty years after the classic proposal by Aaron Temkin Beck (1987), inventor of the famous Beck Depression Inventory as well as a few other decidedly gloomy scales.

Yet, we are still a long way off from understanding how negative bias and depression actually come together in neural circuits. Given the therapeutic effects of drugs that interact with the neurotransmitter serotonin—selective serotonin reuptake inhibitors (of which Prozac is the best known brand)—we have a clear target for research in this area. The topic is finally beginning to surface in fMRI studies (Roiser et al., 2008; Van der Veen et al., 2007). There is learning going on, which I take to be evidence that some stubborn people share with me the belief that there could be some use for it, that learning does matter, that it is the only way in which we can hope to eradicate diseases such as depression.

Eradicate depression, yes, but not fear and sorrow in small and useful fits, for with the romantic poet John Keats, the last of the great six of which William Blake was the first, we return to the theme of the singularity of restlessness, the need for chiaroscuro, the necessity of deviations in opposite directions in order to experience the fullness of life, or life to the fullest. In *Ode on Melancholy*, first published in 1820, in the before-last year of his very short life, Keats put it aphoristically: "Ay, in the very temple of Delight/Veil'd melancholy has her sovran shrine" (1974, p. 201). He saw this not merely as a matter of fact, but as a bit of wisdom to learn from and truly live by—for evidence I present the imperatives in the second strophe of this remarkably wise and melodic poem (p. 200):

But when the melancholy fit shall fall
   Sudden from heaven like a weeping cloud,
That fosters the droop-headed flowers all,
   And hides the green hill in an April shroud;
Then glut thy sorrow on a morning rose,
   Or on the rainbow of the salt sand-wave,
     Or on the wealth of globèd peonies;
Or if thy mistress some rich anger shows,
   Emprison her soft hand, and let her rave,
     And feed deep, deep upon her peerless eyes.

4 The Familiarity Fallacy

Twenty men crossing a bridge,
Into a village,
Are twenty men crossing twenty bridges,
Into twenty villages,
Or one man
Crossing a single bridge into a village.

This is old song
That will not declare itself...

Twenty men crossing a bridge,
Into a village,
Are
Twenty men crossing a bridge
Into a village.

That will not declare itself
Yet is certain as meaning...

Twenty men crossing a bridge, followed or not by a comma as they move into a village, make for a perfect image of difference and repetition in *Metaphors of a Magnifico*, a poem first published in 1923 in *Harmonium*, the remarkably rich debut collection by the iconic modernist poet Wallace Stevens (1984, p. 19). From the apparently arbitrary beginnings of the first two lines, the poem quickly proceeds to another plane of description, from text to metatext, raising one of the most fundamental issues in human thought. Does the case we have before us represent uniqueness and difference ("twenty men crossing twenty bridges"), or can we generalize and recognize the structure of what is repeated ("one man crossing a single bridge")?

All acts of classification absorb new information by bringing it to bear on an extant semantic system. "This is old song/That will not declare itself"—much of the cognitive processing that feeds into these acts of classification occurs below the radar, outside of our stream of consciousness. Information matches, or does not match, with the architecture of the world as we know it, and usually we only become aware of how we classify the objects and events that we witness when we have already classified them. The initial newness of what impinges on our senses is *re*-cognized as an exemplar of something known, something encountered before, familiar already, and more or less understood in likelihood terms that connect it with other features and properties that allow us to expect what will happen next and what we can do to bend the future to our advantage. *Or* the newness is declared to be truly and undeniably new for real, as in never ever seen, completely unknown, and therefore (dangerously or excitingly) unpredictable.

Occasionally we learn to differentiate more precisely. We rewrite our semantic system, and find that an additional item of information ("into a village"), separated by a comma from the main part ("twenty men crossing a bridge"), should actually be regarded as an integral component of the event, no longer to languish on the other side of punctuation. "That will not declare itself/Yet is certain as meaning"—the matching continues to happen below the radar, if slightly differently from now on, based on an updated reference frame. The process may be unconscious, but that does not condemn it to absurdity or chaos. The one thing for sure is that the expansion of familiarity and the increased powers of differentiation are all about defining meaning and making the world sensible.

I cannot help but hear anachronistic echoes of such divergent thinkers as Gilles Deleuze and J. J. Gibson in these lines by Wallace Stevens, forcing interpretations onto me that are likely to be peculiar to my semantic system. I might try to put these private associations into words and communicate them to whoever showed an interest, in an effort to move from idiosyncrasy to a shared experience—a prerogative, or maybe even *the* prerogative, of language. Then I might recount how the idea of twenty men crossing a bridge into a village calls to mind, calls to *my* mind, a collage of visual memories, of a summer festival in Shinto tradition, with men in traditional dress carrying a black-lacquered-wood shrine on long poles lashed together with ropes, the entire procession swaying and hopping to the sounds of flutes and drums. "What could be the purpose of running around with the shrine like that?" I thought aloud one time, and was told that the shrine blesses the neighborhood—marking off territory, much like a dog might, by distributing the scent of self, I thought aloud again and would have wished the thought had never entered my mind, for now I was instantly and quite rightfully the object of vigorous rebuke. Mine was not a pleasant comparison, and did not the fact that I was able to raise it in the first place suggest a sinister bent in my character?

The Majority Rules

"But I really love dogs," I might say, "and I did not mean any disrespect." I would probably make my case only that much worse, refusing to acknowledge my error of judgment and the offense caused, the way the members of the Spanish basketball team at the Olympic Games in Beijing added insult to injury when they failed to own up to their callousness and stupidity, having posed slanting their eyes in what can easily be mistaken for a blatantly racist gesture (e.g., "Spanish basketball players defend controversial photo," CBC. ca, August 13, 2008). Arguably, the inability to anticipate racist readings of our behavior, your behavior, mine, or that of someone else, however innocent in intention, constitutes in and of itself a form of gullibility, or even narrow-mindedness, that could make us end up singing along with the wrong majority at the wrong time and in the wrong place. Sometimes it takes a bit of willingness to think independent thoughts to see what is right and what is wrong. Critical thinking would be the first step in building up the courage to swim against the tide—a tide that can sometimes be pretty ruthless as expressed here in one of the most famous speeches of the last century:

I could easily imagine a life which in itself was fairer than to be hunted through [X], to be persecuted by countless Government regulations, to stand constantly with one foot in gaol, and to have in the State no right which one can call one's own. I could imagine for myself a fairer destiny than that of fighting a battle which at least at the outset was regarded by all as an insane chimera.

These words are taken from one of about a hundred fifty speeches collected in *The Penguin Book of Twentieth-Century Speeches* (MacArthur, 1999, pp. 118–119). I crossed out the country through which the speaker was hunted so as to emphasize the similarity of this case with so many others in the horrible world history of the past hundred years. Was it South Africa, Russia, or India? Are we talking about Mahatma Gandhi, Alexander Solzhenitsyn, or Nelson Mandela?

The three named individuals undoubtedly deserve to be regarded as some of the most honorable specimens of our species, but the words were actually spoken in Germany, at the Industry Club in Düsseldorf. The year was 1932, and Adolf Hitler spoke in front of a group of industrialists, rousing them, warming them up for support of his Nazi movement. It certainly surprised me to find these words spoken by quite possibly the most harmful creature that ever walked the face of the earth, and unless there is something fundamentally wrong with me, I bet it will surprise others who read these words as well. I offer them here in the hope that they may remind us of a thing or two: that the dynamics of majority and minority do not in and of themselves imply justice in one and immorality in the other, for instance, and that we should not be too confident about our ability to spot evil even if it happens right in front of us. We are also reminded that language is an amazingly powerful tool to lead us astray and generate false expectations. Adolf Hitler was a master at it.

In a much more mundane way, I guess I played a few tricks myself in an effort to stage a surprise, carefully picking a misleading quote, bringing it in the context of an argument against racial injustice, and situating it in the company of three true heroes. I deliberately restricted the "answer space," or the range of possibilities in the foreground when contemplating who might have spoken the quoted words—a mechanism that Noam Chomsky (2002) has repeatedly identified in how the media cover politics in the United States, focusing virtually all attention on relatively small differences between Republican and Democratic ways of seeing the world, and effectively limiting the width of the entire political horizon to the little space that exists between the two models.

It is a form of bias that I have touched on already in chapter 1 when addressing the effects of reducing or increasing the number of alternatives in decision making (e.g., Churchland, Kiani, & Shadlen, 2008). The bias in favor of a handful of given options would selectively raise the starting point for the corresponding decision processes, giving these a greater chance of victory than the undefined infinity of all other options. To work against a bias of this kind requires breaking open the answer space, or thinking outside the box—the proverbial hallmark of creativity. Of course, the earlier caveat applies here as well. Creativity is not in and of itself a good thing. The world would surely be a better place without the varieties of creativity that Adolf Hitler displayed.

Other parts of that infamous speech by Hitler in 1932 more readily reveal the nature of the beast. By 1934, he was saying things that could hardly be misunderstood at all. Here is an excerpt of his first public speech after the Night of the Long Knives, or the night he eliminated his rivals within the Nazi party (MacArthur, 1999, p. 134):

I gave the order to shoot those who were the ringleaders in this treason, and I further gave the order to burn down to the raw flesh the ulcers of this poisoning of the wells in our domestic life and of the poisoning of the outside world.

This is a language that, cunningly employing algorithms borrowed from poetry, relishes its power and delights in the use of violence. It is also conspicuously contradictory with the previous excerpt—Hitler now doing exactly (and even more extremely) what he reproached the State for two years earlier. We are likely to fall prey to hindsight bias, though, if we judge these words to be unmistakably those of a monster (see Blank et al., 2008, for a contemporary examination of the cognitive phenomenon, or rather the cognitive phenomen*a*, of hindsight bias). With the inevitable salience of the known future, it may be hard to imagine any real prospect of other futures back then. We are likely to underestimate the level of uncertainty and ambiguity, and the amount of noise, in the minds of listeners at the time Hitler spoke his damnable words.

The context of all that happened since, combined with our inability (or lack of effort) to cross out what we know, makes it easy for us to pass judgment on Hitler's listeners, being so desperately wrong to cheer for the monster. We may wish to believe that we would never have made their mistake, but how can we be sure? How come so many people *did* make the unfathomable mistake of cheering for Hitler? Blaming everything on some kind of mass hysteria, we run the risk of failing to recognize the underlying mechanisms and being governed by these again in a future dark age. Perhaps the more relevant observation about mass convergence onto a single viewpoint, even as deranged as that of Hitler, is that the sheer scale enlargement produces increased salience and familiarity of ideas and, inherently, a facilitated acceptation of whichever opinion is voiced by the majority. "The majority rules" is an honorable democratic principle but no guarantee of fairness. When majority opinion is deliberately shaped by propaganda and propagated by unanimous mass media, we speak of "brainwashing"—one of the oddest and most ironic idioms in English. The effect that it refers to comes down, once again, to limiting the number of alternatives, ideas, or viewpoints, but this time in such a radical way as to leave exactly one almighty singularity of opinion, creating the most extreme familiarity fallacy, urging people to decide that things are as they should be when they chime with what everybody else is chanting.

We are sure to be biased in favor of an alternative if that happens to be the only alternative we can think of. This is not to say that this kind of bias is structurally *wrong*, that it necessarily implies an ethical issue, or that there cannot be such a thing as a single truth to a specific situation. In a variation on Descartes's (2007) skeptic reasoning that produced the first principle of *cogito ergo sum*, I would like to offer the opinion that I exist, and since I can only think of me as a being in existence, I am sure to be biased in favor of thinking that I exist (though I am happy to concede right away that "I" is a pretty ill-defined word, here simply used as shorthand for the subject that proclaims its own existence).

The tricky question, to which my biased me sees no simple answer, is how we can distinguish between familiarity biases that are right and those that are wrong. When do we fail to think of better alternatives because there are none, and when do we fail because we did not exercise our powers of imagination and simply drifted along with erroneous mainstream opinion? Could we employ knowledge of the familiarity fallacy to expose singularities of opinion, which then should be subjected to a round of serious thought and exploration to confirm those very singularities to be right or wrong? It would be an algorithm of sorts, but a tedious one, with plenty of rounds of rethinking, many of which would seem to be extremely debilitating (we cannot keep questioning everything if we are to gain incremental knowledge of anything).

In many ways, the scientific community has explicitly implemented familiarity biases to do away with excessive skepticism and move on with "the real work"—we rely on the expertise of our peers and often choose to take certain things for granted while expending our most critical cortical energies only on a small handful of questions (usually the ones about which we publish research papers). Thus, we look for familiar names—of research institutes, journals, grant agencies, and principal investigators—that have a strong reputation. People who come with credentials and endorsements from well-known corners will more readily be hired for academic jobs. Work that comes from well-known corners will more readily be read and accepted than whatever comes from elsewhere.

It is, of course, one of the reasons why I desperately wanted this book to be published by The MIT Press. In saying so, I admit exploiting familiarity biases and using heuristics based on reputation in my own professional life. I have to—otherwise I would lose too much time, learn less, and end up being unproductive. The familiarity biases are indeed very effective and should in this instance probably be considered a convenient shortcut rather than a fallacy. In the scientific community, the bias mechanisms relying on reputation tend to work almost paradoxically well because the strength of a track record manages to somehow still be a fairly reliable currency, seeing as all of the relevant accolades, publications, grants, and degrees are open to (sometimes rather fierce) competition and subject to scrutiny by peers or superiors. Presumably, the false alarms and erroneous decisions due to the familiarity fallacy would cancel each other out, and any one of us would only manage to build up a convincing record if she or he had some real substance. The scientific community may be as good a model of meritocracy as we can find anywhere in human society.

Relying on familiarity when making decisions, then, may work if familiarity is merely employed as a quick predictor in situations that do not afford lengthy rounds of exploration for better alternatives. However, for biases of this kind to remain efficient, they have to be flexible, corrigible, and adaptive, exactly as in the example of the scientific community, in which the performance of people and institutions is under constant scrutiny. In essence, this involves a hybrid model with a horizontal organization of many decision makers (in this case, scientists) who apply familiarity biases for all but a small portion of decisions but take great care to expose fallacious assumptions for that remaining small portion of decisions, usually within their specific area of expertise. In this way, the statistics of how things go together and which associations are valid are updated continually in the light of new information—a sine qua non if we are to make decisions that stand the test of time.

What we accept as a given, a fact known to be true, trusted, and familiar to us, will fail to give us proper guidance if the expectations are computed on the basis of an outdated, stagnant, or fossilized mental model of the world. This is, of course, exactly what happens under leadership gone awry as we have sadly witnessed countless times in as many dictatorships in recent world history. Many a revolution might have started out with an idealistic component, and some responded to quite obviously reprehensible circumstances, but all too often the destructive energy employed to generate the change failed to translate into a constructive energy that would enable a structure of power open to criticism and ready to correct mistakes.

Strong leadership, relying on its own set of assumptions and familiarity fallacies, might be a winning instrument for the reconfiguration of a political landscape and quickly gain the support of many. Yet, once the deed is done and the new power in place, its pyramidal architecture can make it difficult to leave room for revision of just about anything. In the worst case, leadership becomes entangled with the personality of an individual who assumes the role of enlightened despot, determined to stick with the once-so-useful familiarity biases and prepared to destroy any (perceived) challenge of authority. Freedom of speech is then usually the first human right that people lose on a slippery slope to imprisonment, torture, exploitation, random violence, genocide, and all of the most atrocious behaviors that our species has in its repertoire.

One of the hallmarks of this kind of *Journey to the End of the Night* (to borrow a title by Louis-Ferdinand Céline, a hugely talented writer, but like, for instance, Ezra Pound a notorious fascist and anti-Semite) is how thinking transforms into believing, or probability into certainty. Familiarity biases, and expectations of how things will turn out ("Muslims will not be good American soldiers"), morph from predictions to prescriptions, setting norms of how things should be ("Muslims should not be good American soldiers"). Sooner or later, decisions are made without a single look at any relevant evidence

("Muslims cannot enlist to become American soldiers"). Fortunately, the example about Muslims and good American soldiers is a fictitious one. In his endorsement of presidential candidate Barack Obama on October 19, 2008, former Secretary of State Colin Powell reminded us of the fate and faith of Kareem Rashad Sultan Khan, a twenty-year-old Muslim who sacrificed his life in service of the American people during the Iraq war. Unfortunately, the mere fact that a reminder was needed suggests how dangerously strong the familiarity fallacy about Muslims and good American soldiers has already become.

In a less extreme version, when decisions are still being made on the basis of evidence, familiarity bias leads to working with double standards—I use the common expression of "double standards" deliberately here as a direct allusion to signal detection theory and the LATER model, implying that the threshold for a decision is at a different level relative to the amount of available evidence when the decision is compatible with familiarity bias as opposed to when it is incompatible. Specifically, within the framework of signal detection theory, the criterion would be shifted toward lower levels of neural activity for a familiar signal, creating a comparatively large fraction of false alarms under the noise distribution. For unfamiliar signals, the threshold would be at higher levels of neural activity. Using the terminology of the LATER model, the decision line for familiar signals would start at a position closer to threshold than the decision line for unfamiliar signals. The conceptualizations with signal detection theory and the LATER model both allow us to compute an exact metric for the double standards and to understand exactly how the information processing differs for familiar and unfamiliar signals, or for things that conform to our expectations as opposed to those that go against the grain.

When exposed in clear daylight, such double standards are the stuff that can make honest people recognize their prejudices and rewrite their decision-making processes—I present it as one of the major reasons for hope that we will yet become a species that puts its brain to good use. Here is a classic example, the lawyer Clarence Darrow speaking in his final plea, lasting nearly eight hours, to defend Henry Sweet—the place Detroit, the year 1926 (MacArthur, 1999, p. 97): "You know what this case is. You know why it is. You know that if white men had been fighting their way against colored men, nobody would ever have dreamed of a prosecution." Henry Sweet, a black man, was found not guilty by a jury of twelve white men ("his peers"), who showed the clarity of mind and had the courage to decide that the doctrine that "a man's house is a man's castle" should apply to blacks as well as whites. It was one of the landmark events on the way to greater equality among people in the United States. Some forty years later, Martin Luther King would make a memorable speech at the Lincoln Memorial in Washington, D.C., on the centenary of Abraham Lincoln's Emancipation Proclamation, with such high notes as this one (MacArthur, 1999, p. 328): *"Now* is the time to lift our nation from the quicksands of racial injustice to the solid rock of brotherhood." Yet, even in this righteous and courageous call to end one injustice, it is almost ironic how, unwittingly, another form of discrimination slips in with the use of the word "brotherhood." It took many righteous and courageous calls from feminists in the 1960s and 1970s to expose the implicit sexism abundant in the fabric of society and the matrix of language. Singer Kim Gordon of the alternative rock group Sonic Youth exposed a similar bit of hypocrisy in the album *Fear of a Black Planet* (1990) by the socially conscious hip-hop group Public Enemy. She invited Chuck D, the leader of Public Enemy, to rap a few words in the song "Kool Thing," from Sonic Youth's album *Goo*, released in the same year. "Fear of a female planet?" she asks him in the song, and Chuck D rumbles in return, "Fear of a female planet. Fear baby!"

The learning is ongoing. On June 7, 2008, Hillary Rodham Clinton conceded the Democratic nomination for the presidency in a speech at the National Building Museum in, again, Washington, D.C. "Although we weren't able to shatter that highest, hardest glass ceiling this time," she said, "thanks to you, it's got about 18 million cracks in it," referring to everyone who had voted for her. The "18 million cracks" became an instant idiom in the English language. The "glass ceiling" had been one for some thirty years already, a vivid image of the limited advancement opportunities in society due to unwritten policies of discrimination. A woman as president of the United States is no longer unthinkable.

The way forward, in the direction of a better grounded principle of equality, is that of an expanding circle—*The Expanding Circle*, as argued by possibly the most influential and controversial philosopher of our time. Peter Singer (1983) used the phrase to suggest that ethical principles such as equality and fair treatment should not be intrinsically linked with the human species. Here, we might become aware of other types of discrimination in the fabric of society and the matrix of language, where the word "animal" is often taken to refer only to nonhuman animals, and animal proteins are typically consumed without much agonizing, except in the taboo case of human proteins (labeled "cannibalism," a high-arousal word with a rather peculiar etymology). Some of Singer's ideas sound positively shocking at first, but they may at least serve to sharpen our thinking and make us consider which kinds of inequalities and differential treatments are justifiable within our way of being in the world.

Of course, it is not easy to enlarge the network of what we consider familiar, and consistent with its obvious roots in the word "family," we can expect familiarity to arise from strong innate tendencies toward, or evolved modules for, a skewed distribution of positive passions for people and things that belong to "us" (see figure 4.1). We do not start from a blank slate, and this is actually something we should be happy about (see Pinker, 2002, for a convincing argument). Singer's ideal of the expanding circle probably requires a substantial reorganization of the neural networks that most of us, humans, have developed without explicit effort, if not naturally. Luckily, we can also be happy about reorganization being quite feasible. At least the nurture components of our development can accommodate an enlarged in-group and a more inclusive semantic system in terms of familiarity. History provides the evidence and shows that it usually begins with lucid and learned humans who, consciously and very explicitly, pioneer a different usage of language and stimulate redistribution in the priorities of information processed in perception and memory. Once the new memes take hold and become salient in culture, they are easily and implicitly incorporated by future generations of neonates. Our children will "naturally" be able to imagine a woman serving as president of the United States.

Figure 4.1
Enveloped in the arms of a creature known since birth, the contented child confidently looks into the lens of the camera, exhibiting early evidence of a family-oriented familiarity bias. The subject is the same as the clamoring one in figure 3.3.

Implicit Associations and Flat Taxonomies

However fair or unfair, "the network of what we consider familiar," or the architecture that underlies the familiarity fallacy, propagates information in a nonrandom, unequal way across neural circuits. The activation spreads through preestablished connections among neurons, whether learned or innate, whether culled or cultivated by experience, from or on the basis of anatomical projections given at birth. Ideas of this kind were first formulated several decades ago (Anderson, 1983; Collins & Loftus, 1975), took hold under the name of "connectionism," and allowed psychologists, philosophers, neuroscientists, and researchers in the field of artificial intelligence to find common ground in the study of *How the Mind Works* (to quote another title by Steven Pinker, 1999, famously repudiated by Jerry Fodor, 2002, in *The Mind Doesn't Work That Way*).

In essence, the spread of activation through a network is taken to mean that the stimulation of any one unit (a neuron or an idealized average of all neurons that respond to the same input in the same way) leads to a change in the baseline firing rate of other units in a graded fashion, such that units that are "closely associated" with the stimulated unit show a higher increase in baseline activity than units that live at a larger "semantic distance" from the point of origin of the pulse. Thinking of *dog*, we might quickly arrive at *cat*, *leash*, or *retriever* but leave *snake*, *knife*, and *decoder* at peace. The process of semantic priming here implies that the neural representation of *dog* affects only a fraction of its surroundings. This happens in a way that exhibits the basic neural signatures of bias as I outlined them in chapter 1, in terms of both anticipatory processing and additive scaling, raising, in the case of *dog*, the activity level of units such as those representing *leash* before there is any explicit acknowledgment of the desire to go for a walk.

Greenwald, McGhee, and Schwartz (1998) introduced the Implicit Association Test as one of the ways in which we can probe the architecture of an individual's semantic system, measuring the degree of coactivation among different words or concepts. This is done using a two-choice task in which subjects might be asked to distinguish, for instance, between flowers and insects, and also between pleasant and unpleasant words. The task can generate something of a semantic compatibility effect, showing improved performance when strongly associated categories share a response key (e.g., flowers and pleasant words) as compared to when less associated categories share a key (e.g., insects and pleasant words). Greenwald and colleagues (2002), perhaps too eagerly, derived "a unified theory of implicit attitudes, stereotypes, self-esteem and self-concept" from data obtained with the Implicit Association

Test (see the challenge by Blanton & Jaccard, 2006), but the proliferating literature around the test certainly does suggest that it measures *something*, catching the individual off guard, say, when a self-described unprejudiced white subject turns out to respond more quickly for combinations of white and pleasant than for black and pleasant. In the meantime, the first reports of neural correlates of automatic sexist and racist associations are surfacing in bona fide scientific journals, marrying fMRI research with the Implicit Association Test (e.g., Cunningham et al., 2004a; Knutson et al., 2007).

Until proven guilty, the self-described unprejudiced white subject who gets a damaging report following the Implicit Association Test might, in fact, still be objectively describable as an unprejudiced white subject in real life when making decisions and engaging in all types of actual behavior. There is good reason to make a distinction between the implicit structure of an individual's semantic system and the way in which the semantic system modulates the translation of thoughts into gestures. Indeed, the first fMRI study that employed the Implicit Association Test (Cunningham et al., 2004a), possibly inadvertently, showed a conspicuous limitation of the predictive validity, suggesting that the test correlates well with early automatic aspects of information propagation but less with more deliberate forms of evaluation. Cunningham and colleagues investigated, as they put it, the neural components of automatic and controlled social evaluation by asking white subjects to view photos of black or white faces for a very brief duration (30 milliseconds) or somewhat longer (about half a second). The subjects showed greater activation of the amygdala in response to black faces than to white faces when the images were presented very briefly, but this difference was significantly reduced for the longer stimulus presentation. If the subject was given sufficient time to look at the photo, the frontal cortex showed greater activation for black than for white faces.

Given what we know about the function of the amygdala and the frontal cortex, the data suggested that an automatic emotional pulse in response to a black face was then corrected by a more deliberately controlled rational mechanism. Of interest is that the score on the Implicit Association Test correlated specifically with the size of the first-wave emotional pulse in the amygdala. Could it be that the Implicit Association Test allows us to probe exactly, no more or no less than, the distances and strengths of associations among items in a semantic network? Knowledge of the "spatial layout" of the semantic network would allow us to guess at the built-in biases for a passively moving stream of consciousness, and even to map how the emotions will flow under extreme circumstances when information is brief or hidden and snap decisions are made on the basis of gut feelings in panic situations that appear to demand, or do anyhow elicit, immediate action.

However, the Implicit Association Test in and of itself provides no character study. We should wish to examine the health of a person's frontal cortex before we use the big labels of racism, sexism, or any other form of bigotry—otherwise, we might end up in a decidedly absurd utopia, where we use the Implicit Association Test to preemptively incarcerate future criminals bound to be guilty someday of a heinous act of pride and prejudice, or hate and ignorance, only to find out that, given the latest data, we should lock away just about the entire human population of planet Earth for having a spatially laid out semantic system. (Steven Spielberg's movie *Minority Report* acts out a science-fiction story of this kind, in which even Tom Cruise, or I should say, the character played by Tom Cruise, realizes how deeply erroneous is the vision that inspired the all too eagerly crime-predictive future in which he lives.)

We do well to remember that information propagation, semantic priming, and spreading activation are, at least to some extent, products of learning, and that the ability to learn is an obvious prerequisite for flexible and well-adapted behavior. As before, the difference between good and bad bias, or the question of whether a given familiarity fallacy is really a fallacy, must be studied in context. A major step along the way to developing good biases and avoiding fallacies is to understand the basic and pervasive mechanism of bias—not ignoring or denying it but facing it, working with it. As Scott Plous (1993, p. XV) put it candidly in his preface: "Once you recognize that a certain situation typically leads to biases or errors, you can often sidestep it or take remedial actions." As I wrote at the beginning of this paragraph, the ability to learn is an obvious prerequisite for flexible and well-adapted behavior. The question for those of us who wish to be flexible and well-adapted people will always remain to what extent we are able to learn.

Scott Plous (1993) structured his courageous introduction to *The Psychology of Judgment and Decision Making* around about a dozen heuristics, biases, social and contextual influences, and common traps, borrowing heavily from the works of Daniel Kahneman and Amos Tversky. Such lists can be quite depressing for the same group of those of us who wish to be flexible and well-adapted people—how would we ever be able to keep track of all of the charted biases that we are supposed to sidestep? Would we not in the very act of sidestepping one bias fall in the trap of another? Wikipedia, the online free encyclopedia (consulted on October 28, 2008), includes over a hundred known entities under the "list of cognitive biases," from the "Texas sharpshooter fallacy" to the "Lake Wobegon effect." My reflex in the face of this kind of wild proliferation, endless ramification, and spurious mystification is to try to go back to the basics, think about what it really means to make a decision, and consider how this process can be influenced by what we know (or what we think we know) and what we feel good or bad about.

I turn to signal detection theory and the LATER model to help me understand the principles and the recurrent structure of decision making, making use of a literature that is conspicuously absent in the review by Plous (which has no David Green or John Swets in the author index, not to mention the perhaps more understandable, if not forgivable, omission of any reference to R. H. S. Carpenter or Roger Ratcliff). Rather than devising a flat taxonomy, I would like to understand what characterizes "bias" as a generic type of influence on decision making. Rather than repeating Wikipedia's list here, I decided to apply Miller's (1956) magical number of seven plus or minus two, or the notion that we have some limits in our capacity for processing information.

It had to be possible to write a book about bias in just seven chapters, addressing the whole of the topic in a reader-friendly way, demanding no excessive exploits from your or my memory. And sure enough, I found everything that I know and think and suspect about bias to be compatible with being wrapped in seven chapters with titles that, for me at least, are easy to remember and give a proper hint of what they talk about. This is not a matter of stylistic fetishism, and even less an exercise in self-congratulation. Instead I see it as a fundamental requirement for any treatise that aims to achieve a noticeable change in the way people think about a certain topic. Soundness of architecture and clarity of vision converge to form the vanishing point of ideal science. If I was not ready to aspire toward that point, I think I would have done you as well as me a better favor by never having started writing this book. A flat taxonomy of phenomena can certainly deserve appreciation as a project of systematic description, but it does not give us knowledge in the sense of understanding how things work, how come they do what they do, or why they do not do what they usually do when something is off, up, or down.

The ambition must be to translate or reorganize lists such as those of Plous (1993) and Wikipedia to a leaner framework with vertical as well as horizontal relations, going from formal questions about what bias is and how we can recognize it in numbers (chapter 1, "Bayes and Bias"), to the influences of what we care about, positively (chapter 2, "Wish Come True") or negatively (chapter 3, "Fear Materialized"), and what we think we know (chapter 4, "The Familiarity Fallacy"), which is something influenced by the positive and negative dimensions in our experience and so naturally follows discussions of positively versus negatively oriented bias. The structures of what we know, however, can be thought of as a form of grouping, and other factors, mere physical factors, might also influence grouping of information (chapter 5, "The Proximity Trap"), whereas grouping itself is a recursive process that paradoxically achieves reduction and amplification (chapter 6, "Less Is More"). At the same time, we do well to (re)visit the undercurrent and stimulate the counterculture that allows us to guard against the costs or damaging effects of error in the operation of bias mechanisms (chapter 7, "Utopia—A World without Bias").

This midflight recap, I hope, further illustrates my discursive logic, which aims to be syntactic rather than paratactic—though I can certainly be appreciative of the latter approach in a different context, as it is perhaps more typically poetic, bringing to mind the famously absurd classification of animals invented or reproduced by Jorge Luis Borges in "John Wilkins' Analytical Language" (Borges, 1999, p. 231):

> These ambiguities, redundancies, and deficiencies recall those attributed by Dr. Franz Kuhn to a certain Chinese encyclopedia called the *Heavenly Emporium of Benevolent Knowledge*. In its distant pages it is written that animals are divided into (a) those that belong to the emperor; (b) embalmed ones; (c) those that are trained; (d) suckling pigs; (e) mermaids; (f) fabulous ones; (g) stray dogs; (h) those that are included in this classification; (i) those that tremble as if they were mad; (j) innumerable ones; (k) those drawn with a very fine camel's-hair brush; (l) etcetera; (m) those that have just broken the flower vase; (n) those that at a distance resemble flies.

The list was made famous, in part, by Michel Foucault (1994) when he claimed in his introduction to *The Order of Things* that the passage had inspired his own venture into the archaeology of the human sciences, one of the great classics of twentieth-century philosophy. The core question, for Foucault, for science and human thought, is the structure of knowledge, how it defines what makes sense, and how it changes over time. It may also be the key issue to focus on when we attempt to come to grips with, or build a possible resistance to, the familiarity fallacy. Only by addressing the structure of knowledge, and by acknowledging its historical dimension, can we move from passive recognition ("implicit knowledge"), and a mere intuitive feeling that certain things are associated, toward explicit statements of functions, dynamics, and causal relations among things.

Of course, it is not literally my goal to restructure Wikipedia's list (for one thing, long as it is, I do not think it is exhaustive). I am only claiming that it can be done, or should be doable. Other than that, I will do no more than take arguably the two best known biases—or heuristics, as they are called—and briefly situate them within the framework that I am developing here. My targets are numbers 1 and 2 in Plous's (1993) section on "Heuristics and Biases," that is to say, the representativeness heuristic and the availability heuristic, both proposed by Kahneman and Tversky (representativeness was first discussed in Kahneman & Tversky, 1972; the concept of the availability heuristic entered the public domain with Tversky & Kahneman, 1973).

The representativeness heuristic implies that we judge the probability of things by how well they fit our "mental model" without taking into account baseline probabilities, or without fully relating things to our background knowledge. The textbook example is that of Linda, "31 years old, single, outspoken, and very bright," who "majored in philosophy" and "[a]s a student…was deeply concerned with issues of discrimination and social injustice," and so on (cited in Plous, 1993, p. 110). When subjects are asked which is more likely—"Linda is a bank teller" or "Linda is a bank teller and is active in the feminist movement"—nine in ten give the wrong answer, saying bank teller plus feminist is more likely than bank teller alone (Tversky & Kahneman, 1982).

I would surely be one of those nine in ten and still feel tempted to give the "wrong" answer, as it somehow does not sound like the statistical question that it is supposed to be. Rather I think it comes in the guise of a general probe about, well, how likely it is that a bank teller could be the person described given that she is a feminist versus not a feminist. That would make for a comparison of P("socially inspired bank teller"|"feminist") versus P("socially inspired bank teller"|"no feminist"). Instead, Tversky and Kahneman (1982) wanted us to compare P("socially inspired"|"bank teller") versus P("socially inspired"|"bank teller and feminist").

Could it be that the example is a bit contrived? Or phrased in a slightly tricky way, so that we simply did not receive enough information about what kind of comparison the researchers were looking for? Perhaps the more solid point about the representativeness heuristic is that we tend to go by "the obvious," do not think things through, rely on implicit associations in flat taxonomies, and commit the familiarity fallacy of believing things to be right as they are if they match as closely as possible with what we generally think to be true. The representativeness heuristic, then, must belong to chapter 4 of this book. A quick search on October 29, 2008, via PubMed ("a service of the U.S. National Library of Medicine" that needs no introduction among neuroscientists and should give great pleasure to many people outside neuroscience, with plenty of free links to journal articles) tells me that as yet no one has attempted to connect the representativeness heuristic with anything neural. However, if I am right, we can expect the underlying mechanisms to be similar to those that produce biases as a function of familiarity.

The availability heuristic, on the other hand, implies that we estimate the likelihood of things on the basis of how easy they are to call to mind (Tversky & Kahneman, 1973). Of course, things can be easy to think of for reasons other than their actual statistical probability—because they happened recently or in the neighborhood (two cases governed by proximity, either spatial or temporal, and properly addressed in chapter 5), because they have been hammered in through frequent exposure (a case for chapter 4), or because they carry important emotional implications, so that they can be understood to belong under "Wish Come True" (chapter 2) or "Fear Materialized" (chapter 3). Within the framework of this book, then, the concept of the availability heuristic is broken up or dismembered into nothingness.

To be sure, this is emphatically not a general evaluation of the pioneering work by Kahneman and Tversky. Many other ideas and concepts remain fruitful and can inspire empirical research (e.g., the loss aversion and the framing effect discussed in chapter 3). But with a shift of perspective, and a redrawn landscape, some characteristic features can disappear from view. As with the representativeness heuristic, a quick search via PubMed on October 29, 2008, suggests that the neuroscientific community has not yet been able to think empirically about how the brain might implement the availability heuristic. If I am right, the concept may be too much of a mixed bag, encompassing situations that really are too different in terms of information processing, making it a topic that does not easily afford an fMRI version.

Something Straight from the Top

Being a consequence of evolved or learned preparedness, the familiarity fallacy comes to us essentially as an effect of context, something—or some things—residing in the brain that readies us—or ready us—for the task of picking up information from the environment in a way that could occasionally and inadvertently produce premature conclusions, false alarms, or mental images that do not agree with the pattern of physical information reaching the senses. Using a slightly (and perhaps unjustly) outdated concept, we might speak, in the case of vision, of "extraretinal" influences to gloss over the difficult distinction between voluntary controlled and more automatic or automated processes and simply refer to influences that do not originate from any ongoing "photic" stimulation (to use another antiquated concept that regrettably seems to have disappeared from the literature and used to be used to emphasize the physical nature of the visual stimulation). Instead, extraretinal influences follow from cognitive processes and neural mechanisms that modulate perception from the top down, endogenously, or on the basis of information that does not come from the retina.

The traditional stance, perhaps appropriately called "Cartesian," was that optical or photic information, in the form of electromagnetic waves bouncing off surfaces, chemically altered the shape of the photoreceptors, inducing a cascade of phenomena, via several layers of cells onto the retinal ganglion neurons, and from there onto the lateral geniculate cortex, then the cortex, where again the information would travel further "into the brain" through the intricate hierarchical visual system with its many modules organized roughly from simple to complex. The basic idea here is of information traveling from the bottom up, toward ever finer representations until we recognize objects and associate these with all kinds of semantic features and episodic memories.

It was still the dominant way of thinking about visual information processing when David Marr wrote his scientific testament—the profoundly moving and groundbreaking monograph *Vision* (1982), with its computational approach to neural processing so new that it required a solid defense. The book has resisted aging—a very rare achievement for a contemporary scientific monograph—in part because it combined clarity of thinking with an unmistakable urgency, fully dedicated to integrating functional analysis with searching and testing algorithms that could perform the task of "seeing," or as Marr defined it (p. 31): "Vision is a process that produces from images of the external world a description that is useful to the viewer and not cluttered with irrelevant information." This is a fascinating definition in its radical emphasis on the usefulness of descriptions. It represented a clear departure from the typical "bottom-up" models of perception and insisted that the seeing of objects be up to the task of being in the world, implying selection as well as compression or abstraction of information. Yet the single weakness of the treatise was that it remained oddly unable to actually incorporate mechanisms of semantic priming or voluntary selection ("attention"), the very phenomena of top–down processing that are there to promote the usefulness of descriptions.

Marr (1982, pp. 100–101) even explicitly raised the classic example of a Dalmatian among a messy collection of black blobs against a white background (an image courtesy of R. C. James and found in many a textbook; see Miller, 1999, for a brief description of what comes "straight from the top"). The Dalmatian is often hard to find at first but easily recognized when presented with the image again. It is such a striking demonstration that many viewers will intuitively feel their perception changing from slow and laborious to instant and snap, a dramatically different style of information processing pointing strongly to a reconfiguration of the responsible neural circuit for visual processing. There had to be some kind of emergent process, a nonlinear organization out of a seemingly chaotic input, but for Marr it was outside the scope of *Vision*. Telling as well is that the work by Robert H. Wurtz and Michael E. Goldberg (discussed in chapter 1), replete with selection mechanisms and internal modes of prioritization, remained entirely absent from Marr's (1982) book—there were no discussions of attention or eye movement control. In hindsight, it is slightly ironic that, despite his wonderful intentions and insightful philosophical comments about vision and the usefulness of representations, Marr kept the core of his computational work going with the traditional flow from the bottom up. Perhaps it was hard to abandon the idea that useful representations in vision were strictly crafted on the basis of information that was actually out there in the physical world. Marr knew that the idea was so dead it deserved to die, but somehow the science of vision did not allow him to properly bury it just yet.

Around the same time, Irving Biederman came on the scene as one of the leading researchers on the topic of context effects in object recognition (Biederman, Mezzanotte, & Rabinowitz, 1982; see Biederman, 1987, for the theoretical framework under the label of recognition-by-components theory). In the 1982 paper, Biederman and his colleagues showed data that looked to be most convincing, suggesting that scene context facilitated the perception of objects that are semantically consistent with, or probable in, the portrayed scene (say a chicken in a farmyard). The experimental paradigm had subjects look for a particular target object, instructed via a target label presented before the actual stimulus, which consisted of a line drawing of a natural scene with or without the object in question. The line drawing was presented very briefly and followed by a location cue. Subjects had to decide, yes or no, whether the target object had, in fact, appeared at the cued location. This turned out to be easier to do for objects that were semantically consistent with the scene as opposed to those that looked slightly surreal, or at the very least a touch idiosyncratic, in their environment (say, a television in a farmyard).

The advantage for semantically consistent objects was obtained with several measures, including percentage correct performance, reaction time, and sensitivity, or d' (pronounced "d prime," a measure borrowed from signal detection theory, here relating percentage correct performance to percentage false alarms). Expanding on criticisms first voiced by De Graef, Christiaens, and d'Ydewalle (1990), Hollingworth and Henderson (1998) worked out one of my favorite efforts of deconstruction in the cognitive sciences, though I guess some of my enthusiasm for this paper betrays a type of bias that I will discuss in chapter 5 (the research blossomed in front of my eyes in the lab where I was working at the time, so I think I stepped right into the proximity trap). The key prize was the sensitivity measure, and Hollingworth and Henderson convincingly argued, with commonsense logic and hard data, that Biederman et al. (1982) had misunderstood their own work. The presumed effects of sensitivity were solely and simply due to response bias. I discuss it in some detail here because I see it as an excellent example of how questions of bias and sensitivity are crucial, and require careful consideration, in the study of perception, recognition, and information processing in general.

The issue centered on the way in which hit trials (with correct detection of the named object) were compared with false alarms on catch trials (incorrectly reporting the presence of the named object). Subjects were, in fact, more likely to make a false alarm when the target label was consistent with the scene than when it was inconsistent. When in doubt, or vaguely having seen something fuzzy, subjects might erroneously convince themselves that, yes, the blobby thing over there must have been a chicken in the farmyard. They were less inclined to falsely project the unlikely television in the farmyard. Biederman and colleagues (1982) fully acknowledged this contextual effect on the likelihood of making a false alarm, but they then went on to use a mixture of semantically consistent and inconsistent false alarm rates to compute the sensitivity measure. This was a mistake—or even several mistakes in one.

First, it violated the basic rationale of signal detection theory, which requires that the correct detection of a particular signal, when it is present, be compared with the false detection of the same signal, when it is absent. That is, the same decision task should be evaluated in different situations (signal present or absent). In contrast, Biederman et al. (1982) computed the sensitivity measure on the basis of correct detection performance versus false alarm rates with different target labels. This also meant that the actual search strategies (e.g., the usage of critical features, or expectations about size and position of the target object) might have been different across normal trials and catch trials, leaving room for an untold number of confounding factors in terms of cognitive processing.

Finally, and perhaps most damagingly, Biederman et al. (1982) computed false alarm rates on the basis of a mixture of trials with semantically consistent versus inconsistent labels. As a consequence, they overestimated sensitivity in semantically consistent trials. This is because the high correct detection rate for semantically consistent trials is now compared with a reduced false alarm rate, given the inclusion of semantically inconsistent catch trials and their relatively low false alarm rate. Vice versa, sensitivity was underestimated in semantically inconsistent trials due to the comparison of the correct detection rate in semantically inconsistent trials with an artificially increased false alarm rate, skewed by the inclusion of high false alarm rates from semantically consistent catch trials...

In short, the flawed analysis of Biederman et al. (1982) accidentally worked in favor of the conclusion that scene context facilitates the recognition of semantically consistent objects. Hollingworth and Henderson (1998) mercilessly and very lucidly exposed the errors and even applied the erroneous logic in their own experiment 1 to prove they could reproduce the results by Biederman et al. (1982). However, with appropriate comparisons, and further changes to improve the object-detection paradigm, Hollingworth and Henderson (1998) provided clear evidence in their experiments 2, 3, and 4 that there was no such thing as improved perception for objects that are semantically consistent with scene context.

Biederman had to acknowledge his old idea needed revision, and he probably did, as he surely must have been one of the reviewers for Hollingworth and Henderson's (1998) paper in *Journal of Experimental Psychology: General*. I find the dialectics between the two papers very exciting—a perfect example of how science moves on and creativity emerges through mutual interaction. The topic itself, then, is all but settled, providing excellent opportunities for further investigative dynamics, I would hope. For one thing, Hollingworth and Henderson were oddly dismissive of the usefulness of eye movements as a dependent measure and did not even analyze them in their paper. With the proposals from the LATER model, it should be perfectly possible to check whether the reaction-time data support the conclusion that any contextual effect favoring semantically consistent objects in the object-detection paradigm is due to response bias. Another notable point is that Hollingworth and Henderson observed something of an "inconsistent object advantage" under some conditions, a slightly counterintuitive idea at first but one that matches well with eye-tracking studies that have shown subjects to stay focused particularly long on unexpected or weird information, seemingly scrutinizing the strange item for the purpose of making sense out of what looks absurd at first (e.g., De Graef et al., 1990; Henderson, Weeks, & Hollingworth, 1999).

At the risk of sounding tautological, or falling for some kind of circular reasoning, we could propose that "informativeness" correlates positively with the extent of information processing, no matter whether the "attention" allocated to the informative information in question be covert (implying internal selection processes) or overt (entailing behavioral orienting responses, that is, eye or head movements in the case of vision). Experimental psychologists typically avoid the tautology by trying to devise alternate means, or convergent methods, to talk about the same thing. One strategy, for instance, would be to establish the informativeness of things via a rating in a questionnaire with one group of subjects before going on to track the eyes of another group of subjects while these latter subjects are picture viewing. This is precisely what Mackworth and Morandi (1967) and Antes (1974) did in their research. Whether "of course" or not, the data proved to be in agreement with the apparent tautology.

With respect to the "inconsistent object advantage," we can then try to think that whatever does not fit in a certain context would be particularly "informative," perhaps because it might require a reevaluation of the gist of the scene or even a little paradigm shift in our interpretation of what it is we have before our eyes. Here, the increased level of information processing would imply heightened sensitivity, an improved signal-to-noise ratio, and the making of more detailed perceptual and memorial representations. Yet, for an effect of this kind to be observed, scene context must be established already before the object can reasonably be said to have the status of semantic inconsistency.

Could it be that semantically consistent items are useful or "informative" at an early stage of information processing, during the construction of a contextual scheme, whereas inconsistent items only acquire a special status once there is a fully established semantic context? Support for this idea arrives from very different corners, including effects of semantic consistency in a behavioral study on "exogenous orienting" (Stolz, 1996) and effects of multiplicative scaling, measured in the neural firing rates of monkey inferior temporal cortex neurons, favoring a target object among familiar distractors (Mruczek & Sheinberg, 2007).

If I read the literature correctly, there would be roughly a two-stage process, one in which a context representation is generated, based on sensitivity mechanisms that work partially from the bottom up, with local features, and partially interact with top–down information, prioritizing semantically consistent information. Once established, scene context would allow the viewer, or subject, or decision maker, to employ bias mechanisms that favor, again, semantically consistent information (see Torralba et al., 2006, for a Bayesian model that already implements some of these ideas). At the same time, though, semantically inconsistent objects would gain a special status as "news," being more informative, potentially a powerful teaching signal in opposition to the established contextual scheme. Items that resist integration with the contextual scheme, then, would elicit more extensive information processing, leading to the paradox of improved sensitivity for semantically inconsistent information.

A Rose Is a Rose Is a Rose Is a Rose

At home, we know where to look for things. The rose which is in the vase on the table by the window must be exactly equal to, and nothing other than, the rose in the vase on the table by the window. The rose by the window on the table in the vase would be something else altogether. At least, that is how it goes according to the logic of the Dutch poet Hans Faverey (2004, p. 40), who might well have been examining the pattern of eye movements when he considered the grammar of (not roses but) chrysanthemums in a familiar space.

At home, we follow fixed sequences when we look for things. Perhaps it is as simple as that. From minimal information, we reactivate a mental map that allows us to make anticipatory eye movements to where we expect familiar things to be. Monkeys do it, too (Miyashita et al., 1996), activating oculomotor sequences presumably coded by a neural circuit that includes a prominent role for the supplementary eye field (Lu, Matsuzawa, & Hikosaka, 2002). As a behavioral strategy in service of perception, anticipatory eye movements manage to perfectly align bias and sensitivity mechanisms, making sure that both types of mechanisms are focused on the same location in the visual field. The prospective nature of the movement brings with it a degree of additive scaling, indicative of bias. But the fact that retinal processing is now centered on the favored location naturally implies that the perceptual description will be finer and more accurate there than anywhere else in the visual field.

Context, the familiarity fallacy, and anticipatory eye movements might even conspire to close the loop, moving from the top down and then back thanks to the information streaming in via the retina, literally providing the objects of thought in an entrained way of thinking, or as the poet Arkadii Dragomoshchenko (1993, first paragraph) intuited: "Habits of mind result from a redistribution of the places on which the eyes fall. Yes, I'm probably right about this." The effect of expertise—the effect of knowing how to recognize an obscure kind of cancer from looking at an X-ray, say, or of being immediately able to spot a tricky kind of offense in a game of chess—would imply well-greased, reliably reproducible, strategic and minimalist modes of visual exploration, requiring just a few well-aimed eye movements (Sharot et al., 2008), and extracting maximal information from these, both globally, addressing the entire spatial configuration of a stimulus, and locally, relying on characteristic object features (Gauthier et al., 1998; see Gauthier et al., 1999, for a potential neural correlate of the role of expertise in holistic processing, obtained with BOLD responses in the middle fusiform gyrus in the so-called face area).

Expertise leads to a peculiar way of seeing, and this way of seeing generates the thoughts that habitually flow in the stream of consciousness. If the proposal appears a trifle complacent and bit too convenient, we should note that it comes with an exciting corollary in the sense that it offers an ideal ground for new figures to emerge. It sets the stage for a sharpened ability to make discoveries. Figure 4.2 invites you to experience the phenomenon firsthand, with a variation of the type of visual search task first investigated by Wang, Cavanagh, and Green (1994). Do you notice anything unusual among these thirteen identical upside-down images of the Mona Lisa? What if you turn the book so that da Vinci's inescapable landmark regains its canonical orientation?

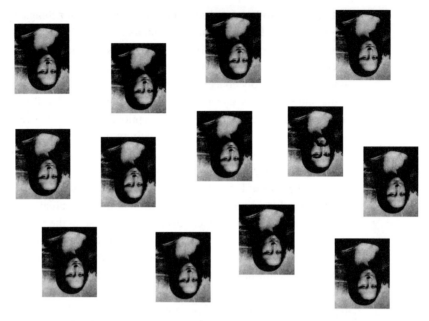

Figure 4.2
Are these thirteen objects identical, or do you detect an odd one out? Turn the book upside down so that the Mona Lisa looks more familiar. Now do you find an odd one out? If your ability to search visually resembles that of the average subject in the study by Wang, Cavanagh, and Green (1994), you will in the latter case have an easier time finding the Mona Lisa with the rather-too-much facial hair (inspired in part by the early modernist artist Marcel Duchamp's treatment of the same overly familiar painting in 1919 and in part by the alternative-rock icon Nick Cave's imago in 2006).

If the little experiment succeeded, you will have more readily noticed the Mona Lisa with the extra facial hair among the twelve normal ones in their canonical orientation. Somehow the background of twelve extremely familiar items enhances our ability to navigate toward the odd one out. In optimal conditions, and with perfectly crafted stimuli, Wang, Cavanagh, and Green (1994) were able to show that the visual search for an unfamiliar item among familiar items can proceed in parallel, with search times that are independent of the number of distractors in the display. In contrast, the search for a familiar target in a field of unfamiliar noise is slow going and error prone.

In all likelihood, the homogeneity of familiar items affords a type of grouping by similarity that is facilitated by a top–down mechanism, a type of semantic priming, not over time but across space. This, I think, must be a bias mechanism, working from the assumption that all items in the set are in fact the same until proven guilty of being different. It implies that the visual system quickly generates very precise predictions for all information in the display—predictions that can then relatively easily be proved wrong if a salient feature mismatches with the familiar template. In the case of unfamiliar items, like the Mona Lisa in an unusual orientation, there would be no semantic priming to facilitate the grouping by similarity. For relatively complex stimuli, such as painted portraits, we may expect, in accordance with feature-integration theory, that this requires sequential processing—each inverted Mona Lisa in turn.

Novelty, then, pops out in a familiar environment. Perhaps this is akin to what in popular accounts of scientific discovery is called "serendipity," when, in the context of an already-familiar future dictated by the projections of systematic investigation, suddenly an unpredicted entity of a new order emerges in defiance of current hypotheses and theories. Here, the level of detail and precision of expectation actually allow the novelty to break through more forcefully.

In a different discourse, discussing the aesthetic principles in the appreciation of photography, Roland Barthes (2000) made a distinction between "studium" and "punctum" that, I think, neatly captures, and further develops, ideas that are similar to those about context, familiarity, and novelty pop-out. Barthes suggested that any image, and by extension any work of art, or state of things, can be said to have a general gist and elicit a dominant interpretation. This is the studium, giving us meanings that are nameable, complete with a set of expectations and associations. However, this is not sufficient for a full (emotional) engagement with the object in question. For that, we need a punctum, a detail that goes against the grain, an element out of order, something unknown, as yet unnamed, something that causes a rupture, appears to negate the studium, and thus creates a tension and urges our cognitive powers to go beneath the surface of the image, to think it through, to contemplate it more carefully, that is, to recruit sensitivity mechanisms.

Moving further down the path of punctum (surely an "event") and its power to restructure the situation ("being"), we encounter what must be the most influential voice in continental philosophy today. I am referring to Alain Badiou, of course, author of the colossal *Being and Event* (2007), who presented his core ideas succinctly in *Infinite Thought* (2005, p. 46):

> For the process of a truth to begin, something must happen. What there already is—the situation of knowledge as such—generates nothing other than repetition. For a truth to affirm its newness, there must be a supplement. This supplement is committed to chance. It is unpredictable, incalculable. It is beyond what is. I call it an event. A truth thus appears, in its newness, because an eventual supplement interrupts repetition.

Here, Badiou applies the logic of studium and punctum to the highest level of abstraction, the ultimate and ontological, *is* and *is-not*. Badiou's is certainly a very deep and stimulating analysis, but I would like to reapply it to the more specific level of human information processing and decision making. Reading things as they are, we obtain a newsworthy event in what is different from the previous "state space" or set of known parameters. Information resides in difference among sameness, in how a figure is different from a background ruled by repetition.

It reminds me of a little (unpublished) observation Ruth Ann Atchley made not so long ago, when she was teaching my colleagues, our students, and me how to use our new equipment for electroencephalography in Wellington. Ruth Ann, an established expert in event-related potential (ERP) research, remarked how virtually all of the well-known ERP or brain wave components were somehow about detecting the oddball, the thing that was off, that did not fit or was different. News was what was new. Badiou, I imagine, would certainly appreciate the poetry of a concept that related a potential to an event—a potential, a power, a bit of electric energy brought about by a truth that affirms its newness.

For Badiou as well as for Barthes, the appearance of an event or punctum represents the moment at which something as yet unnamed demands the invention of a new name—a poetic act, according to Badiou, or simply a matter of perceptual learning, according to J. J. Gibson (1950, p. 222) in *The Perception of the Visual World*:

Objects, events, and situations are recognized. In the case of human beings, things are named. The qualities of, or differences between, objects are also named. This enables us to name classes of objects. Once this process is started it builds upon itself; new differences emerge, new similarities become visible, and more classes are named. At the same time more and more objects are identified. The traditional way of putting it is to say that things have meaning and that we have abstract ideas about them. But the progress of learning is from indefinite to definite, not from sensation to perception. We do not learn to have percepts but to differentiate them. It is this sense in which we learn to see.

We adapt to the event of the new by giving it a name and using the characteristic difference as a diagnostic feature in perception. Once named and a new class of objects recognized, the sameness prevails again, a new kind of familiarity fallacy emerges for the freshly baptized category, bringing along its own variety of bias in the principally not so crazy assumption that things will be as they most likely are.

The ability to learn and to expand our vocabulary, I hope, will continue to allow us to put the familiarity fallacy to good use, avoiding its negative potential for bigotry and prejudice. As I am wrapping up this chapter 4, I cannot resist emphasizing the highly optimistic frame of mind in which I happened to write these thoughts. Just two days ago, Barack Obama was elected to become the first African American president of the United States. With the event comes a new name, President Barack Obama, and with the new name, and our increased ability to differentiate among percepts, we will replace old expectations of heterosexual white male presidents with new ones, "young and old, rich and poor, Democrat and Republican, black, white, Hispanic, Asian, Native-American, gay, straight, disabled, and not-disabled," to quote a list given by the president-elect during his victory speech at Grant Park in Chicago on November 4, 2008 (he forgot to mention male or female, but I think we can take it for granted from now on that our mothers, sisters, and daughters are fully included in the list).

From difference to repetition and back, we move. With the grouping by similarity, or the being in the in-group of the familiar, we are already headed for the proximity trap, or the topic of Gestalt psychology and perceptual organization, the main theme of chapter 5. The grouping by familiarity, nevertheless, stands apart as a type of perceptual organization that depends on learning, works from the top down, and naturally moves into the sphere of language (and power, and politics—but first of all, language). And so it is we can suppose that a rose is a rose many times only to find out that sometimes there arises here or there something new, something that will come again. In conclusion, let me quote the actual verbal choreography of Gertrude Stein (2008, p. 186) in "First example" from *An Elucidation*, written in 1923:

Suppose, to suppose, suppose a rose is a rose is a rose is a rose.
   To suppose, we suppose that there arose here and there that here and there there arose an instance of knowing that there are here and there that there are there that they will prepare, that they do care to come again. Are they to come again.

5 The Proximity Trap

```
so
]
]
]
]
]
Go              [
so we may see   [
]
lady
of gold arms    [
]
]
doom
]
```

This is not a piece of radical vanguard poetry, deliberately bordering on the incomprehensible, but all that remains of a poem by Sappho, the great Greek poet who sang her words to the lyre more than 2,500 years ago. The music is lost, and of her nine books of lyrics a total of exactly one poem survived intact. Other than that, we have a considerable amount of bits and pieces quoted by others, or found among precious scraps of papyrus. The American contemporary poet Anne Carson scrupulously examined and translated the fragments to produce an eerily beautiful volume, titled *If Not, Winter*. I picked the fragment reproduced above (Carson, 2003, p. 15) for several reasons—singularity of reason never being sufficient cause for true excitement—but I will only mention two, leaving the others open to the pleasure of independent discovery (or closed to the laziness of passive reading). The first reason is the wonderful logic of chance. Somehow the unintentional ellipsis has produced a beautiful little poem in its own right, something altogether different from what Sappho wrote, allowing us to see her work in a way that she could not.

(There are many technical and semantic reasons why I think what we get to see here in our shifted perspective is actually *beautiful*, but I was not going to expand on those.) The second reason belongs to the surface and the actual theme of this chapter 5. The fragment stands as a concrete reminder of the ravages of time. It makes the loss painfully visible and forces us to realize the contingency of our own perspective—however much we may wish to escape the boundaries of time and space, the fact remains that proximity of things to Here and Now has a huge impact on what we notice, remember, prefer, and decide.

The Cheerleader Here and Now

The notion that objects look smaller as they are further removed from us, obvious though it may seem, did not fully enter the repertoire of visual artists in Europe until the Italian Renaissance. Giotto di Bondone made the first gestures toward a realistic sense of three-dimensional space on two-dimensional surfaces in the beginning of the fourteenth century in Florence. It took several more bright minds, including those of Filippo Brunelleschi, Leone Battista Alberti, and Piero della Francesca, and a good understanding of the algebraic underpinnings, before perspective became one of the standard tools in art. Until then, the size of objects was determined as much by their symbolic weight or importance as by their appearance in a visual scene.

Perhaps it is not so much the notion of faraway objects' being small that poses difficulties to the exercise of translating images onto paper. Figure 5.1, produced by a six-year-old child without any training in perspective, clearly incorporates that idea (as well as the modulation of size by symbolic weight, here with an almost disarmingly unashamed self-centeredness). It is rather the systematic and parametric nature of the relationship between distance and size that does not come naturally to our mind. Even if we know it is bound to be operative somehow, we cannot easily see through the proximity trap. (With a bit of reverse etymology: "to see through" in Latin makes *perspicere*, the ancestor of perspective—so we lack perspective on the way the proximity trap plays in perspective.)

In many cases, we will simply fail to be aware of the proximity trap, even when we make explicit attempts to draw a representative map of a state of things in a given domain. We may refer to a disproportionately high number of recent research papers not because they are better or more informative than earlier studies but because they are fresh in memory or because we have first-hand knowledge of how they came about—perhaps we reviewed the paper for a journal, or maybe we discussed it with one of the authors at a conference.

Figure 5.1
Drawing of a school festival, featuring an army of cheerleaders. The giant in the middle represents a self-portrait of the artist against a background of ant-sized classmates. The artist is the same as the clamoring subject in figure 3.3 and the contented one in figure 4.1.

This is not to say that firsthand knowledge should be frowned upon (on the contrary, sometimes it really facilitates crystallization in memory and enables the creation of a landmark in our thinking as happened for me with the 1998 paper by Hollinworth and Henderson, discussed in chapter 4). However, it does become a problem when we let laziness reign and make a habit of processing only our immediate surrounds.

Unchecked, the proximity trap can produce rather severely distorted surveys and reviews. For chapter 4, I was relying fairly extensively on Brian MacArthur's (1999) compilation of *The Penguin Book of Twentieth-Century Speeches*, a book that by its title clearly lays claim to comprehensiveness. The collection received a favorable review in *The Times* as according to the back cover of the book—we are told "[i]t would be hard to do better than MacArthur's selection, which is a tribute to the breadth of his knowledge." Considering that MacArthur is (or was, at the time of publication) an associate editor of *The Times*, we note with a little sigh of relief that the quote in question was due not to the man himself but to one of his colleagues, Charles Powell. MacArthur, based in London, picked 155 twentieth-century speeches, of which more than 120 were spoken in English. Seventy speeches were given by citizens of the United Kingdom (compared to forty-five for the United States). David Lloyd George (a name I had never come across before) turned out to be the author of five (or well over 3%) of the most important speeches in the twentieth century. Aneurin Bevan and Neil Kinnock (a name that was actually immediately recognizable to me) each got three to their credit.

I found no speeches by Mao Zedong, the Dalai Lama, Lech Walesa, Mikhail Gorbachev, Yasser Arafat, or Desmond Tutu. There was not a single Japanese speech, not even Emperor Hirohito's Jewel Voice Broadcast of August 15, 1945, the unforgettably brave and slightly awkward surrender speech, the first time in history an Emperor of Japan spoke to the common people, in which he mentioned that "the enemy has begun to employ a new and most cruel bomb, the power of which to do damage is, indeed, incalculable, taking the toll of many innocent lives" (Montefiore, 2008, p. 116) and further offered that "it is according to the dictates of time and fate that we have resolved to pave the way for a grand peace for all the generations to come by enduring the unendurable and suffering what is unsufferable" (p. 117).

Of course, we cannot forgive the oversight. But perhaps we can a show an ounce of mellowness in our understanding, on the assumption that MacArthur would, upon being confronted with the omission, realize his mistake and correct it in a revised edition. The inability to circumvent the proximity trap is certainly not MacArthur's alone, and it may be much more pervasive in our day-to-day dealings than we would be ready to admit. One extreme example, incisively funny and unsettling at once, comes from a study by Johansson and colleagues (2005), involving a beauty contest and the switching of candidates. The researchers asked subjects in the lab to indicate the more attractive face in a pair of photographs and then handed the chosen photograph to the subjects for a bout of free introspection on why they had chosen that particular face. With a little sleight of hand and a straight face, the researchers occasionally handed subjects not the chosen but the rejected photograph.

Johansson et al. (2005) manipulated the deliberation time (the length of time given to the subjects to make up their mind about which face they preferred) and the similarity between the two faces of a pair. With up to five seconds of deliberation time, subjects failed to detect the mismatch in a staggering 80% of the cases. Given free deliberation time, subjects still failed to notice the trickery in about half the cases. The similarity of the faces hardly had a measurable impact on the detection rate. Striking as well is the content of the subjects' introspections, explaining why they had "chosen" the face that, in truth, they had rejected. Roughly a third of these confabulations referred to a detail or something specific to the photograph (e.g., a male subject noting, "[s]he's radiant. I would rather have approached her at a bar than the other one. I like earrings!" or a female subject noting, "[s]he looks like an aunt of mine I think, and she seems nicer than the other one").

I am not sure whether the formality of the lab situation contributed to the astonishing results. Subjects might have been overly gullible and trusting in the researchers' intentions—on the other hand, there would be something to say for the thought that subjects come to a psychology lab with heightened suspicion that "those people are bound to try something funny." In any case, the salience of the proximal stimulus, the photograph in hand, whether enhanced by the situation or not, sufficed to create a response bias with retrospective power. Fully focused on what apparently had been "their choice," subjects engaged in what Daniel C. Dennett (1991) would have called a Stalinist rewrite of history—contradictory evidence or whatever fleeting trace of it remained in memory was suppressed and washed over by new perceptual descriptions, heavily biased toward the conclusion that the once-rejected face was in fact "attractive." Subjects were now given nothing but the new face to look at—a situation that gave free rein to a gaze bias of the type described by Shimojo and colleagues (2003), implying a formative influence on preference.

The finding resonates with other evidence suggesting that true (but essentially irrelevant) photographs or extra perceptual detail here and now can invigorate a response bias in favor of "remembering" a childhood event that never took place (Lindsay et al., 2004; see also Gonsalves et al., 2004, on the role of vivid imagining in false remembering). Make-believe memories may in general be due to contamination by postevent suggestion, the proximal context overpowering the distal (Loftus, 2003). There would probably be very few false memories indeed if it were not for the proximity trap. The immediacy of the present provides a forceful framework for the (re)shaping of representations in false memories in a way that clearly deviates from that of true remembering. Illusory memories are supported by activity in frontoparietal circuits (perfectly placed for response bias), whereas veridical memories draw their episodic information straight from the medial temporal lobe (Kim & Cabeza, 2007; see also Garoff-Eaton, Kensinger, & Schacter, 2007). Subjects may be confident of the truth of their assertions in both cases, but the underlying mechanisms are vastly different.

Another line of investigation, generally referred to as research on "change blindness" (reviewed by Simons & Rensink, 2005), shows the proximity trap in operation when the saliency of the current image dominates to the extent that an actual comparison with even the immediate past becomes virtually impossible. Even drastic changes often go unnoticed—I remember one striking example, from a talk by the psychophysicist Paul Atchley, with an entire apartment block (standing tall in the background) disappearing from view before I knew it (I think the example must have been derived from the stimulus set of Pringle et al., 2001). Change blindness belongs to a set of phenomena that bring into focus the negative aspects of selective information processing—whatever falls outside the selected window quickly evaporates or vanishes in the dark. In relation to the proximity trap, however, the point that I wish to stress is the unduly large influence of the current image on our understanding of the situation.

Our perceptual system seems to incorporate an expectation of stability, so that we habitually take it for granted that the way things are right now is the way things were a moment ago. From the perspective of neural processing, this is a perfectly viable minimalist assumption, requiring no laborious effort in processing the huge amount of information bombarding our senses at any one point in time. It has the further benefit of enabling a strategic concentration on a subset of the information, where changes are likely or potentially relevant or meaningful in terms of what invites approach or urges avoidance (I will expand on the implications of selective information processing in chapter 6). In one sense, the minimalist assumption of stability actually capitalizes on proximity and appears to even exaggerate its distortion. The here and now prevails in a way that might have been biologically advantageous, in terms of natural selection for our ancestral species, but warrants a detached examination if we are to understand, and possibly correct, our biases.

The process of comprehension, in the case of the proximity trap, begins with the acknowledgment of the limits of any individual perspective, grounded geographically, historically, and culturally. In the domain of visual object recognition, the role of viewpoint was the focus of heated debate a decade or two ago, with Biederman's (1987) recognition-by-components theory and the "geon structural description approach" implying a strictly geometrical kind of viewpoint-invariant processing (Biederman & Gerhardstein, 1993), whereas Michael Tarr and colleagues propounded a "multiple-views approach," with object recognition as a matter of matching perceptual descriptions with stored views of objects (Hayward & Tarr, 1997). BOLD activity during object recognition pointed to a crucial dependency on viewpoint, apparently vindicating the multiple-views approach (Gauthier et al., 2002).

I still think the recognition-by-components theory has computational merits that were never quite superseded by the multiple-views approach, so I look forward to seeing an ambitious hybrid model arise in this area. However, the point about viewpoint dependency deserves to resound for a good while to come. One thing that I find particularly promising about the multiple-views approach is that it establishes a connection between viewpoint dependency and a process of matching on the basis of similarity—or how the proximity trap is not only about here and now, or distances in time and space, but also about more abstract, multidimensional distances (or differences) relating to the shape of a thing. It represents a return of the eternal return, or the issue of similarity and repetition. Here, it is situated at the heart of recognition, of grouping one thing with another.

Grouping by An-Pan-Man

Still an indispensable chapter in any introduction to psychology, the work of Gestalt psychologists like Max Wertheimer, Kurt Koffka, and Wolfgang Köhler drastically altered the landscape of experimental psychology in the beginning of the twentieth century, with its emphasis on the mind's tendency to create objects, shapes, or figures in perception and cognition. The Gestalt psychologists took their cue from philosophical musings by thinkers like Franz Brentano who discussed the "intentionality" of thinking, using the term in a slightly unusual way, meaning something akin to the neologism "aboutness" (Dennett, 1989), referring to the fact that thinking is always *about something*. Thinking necessarily takes an object, even if Descartes's formula seemed to forget that for a moment, with its thinking, and therefore being. Normally we expect some kind of *that* or object phrase to follow "I think." For the Gestalt psychologists, the tendency to focus on *something*—an object or unit of thought—represented a fundamental property of the way in which we process information in the environment.

The gestalts that we construct in perception and cognition serve to organize the available sensory data in tightly packed units that make it easy to handle the situation, literally so, providing us with targets for our actions. The fundamental principle operative in perceptual organization would be "Prägnanz" (or "conciseness"), integrating information as simply or as neatly as possible. Under the banner of Prägnanz come several laws, such as those of similarity, common fate, closure, continuity, symmetry, and, indeed, proximity. Each of these laws allows us to group items in a display, or elements in a soundtrack, on the basis of low-level physical features of the stimuli that activate our senses—precisely defined in spatial and temporal dynamics, purely geometrically, in a way that should relatively easily afford machine learning (see Roelfsema, 2006, for a contemporary introduction in terms of "cortical algorithms").

In real life, it is, of course, not always that straightforward, as some of these laws might, in a given case, pull in opposite directions. Figure 5.2 shows the rather idiosyncratic case of "grouping by An-Pan-Man," in which a

Figure 5.2
The epoch-making case of "grouping by An-Pan-Man." The one-year-old grinning at the camera relishes the resemblance with the cartoon figure portrayed on the front of the stroller provided by a shopping mall in Tokyo, Japan (the resemblance is further accentuated in the original version in color, with the refracted wavelengths of the cartoon figure matching those of the passenger). The cartoon figure, called "An-Pan-Man," is an anthropomorphized red-bean pastry turned superman. The one-year-old is a sibling of the main subject in figures 3.3, 4.1, and 5.1.

one-year-old feels he shares a bond with a cartoon figure that shows a rather remote similarity—a similarity that humans may or may not pick up from among the many possible ways to organize the image perceptually, depending perhaps most of all on their willingness to play along, but the tenuous similarity would surely be hard to pick up for any machine vision program, however hard it wished to play along. Here, familiarity is presumably a decisive factor that helps shape the shapes, of both the one-year-old and the cartoon figure, that then become the object of grouping (the "grouping by An-Pan-Man" might for this reason happen first and foremost, or even exclusively, in the eye of the beholding parent).

I am inclined to think that the ability to pick up surprising similarities or idiosyncratic connections among things is exaggerated in humans. It is characteristically poetic, in a sense that does not limit "the poetic" to the works of poets but rather embraces the etymology of the word "poetry," with its roots in the Greek verb *poiein* (to make or to create). The forging of linkages and the seeing of never-seen similarities possibly depends on, and is definitely facilitated by, language with its modes of abstraction and its devices for imagination, from metaphor to metonymy. But also in the more mundane forms of perceptual grouping, driven entirely from the bottom up, reading nothing but the ad hoc structural properties of the sensory data, we can expect grouping by similarity to be due to reactivation of the same neural circuit, as shown in an fMRI study with unfamiliar objects, in which similar shapes produced similar BOLD response patterns in object-selective cortex (Op de Beeck, Torfs, & Wagemans, 2008). In fact, I suspect that the initial activation of a neural circuit by incoming sensory information relating to a particular object carries with it an implicit bias in the processing of subsequent information, lowering the threshold (or raising the starting point, as according to the LATER model) for recognition of the next object as "the same." The very act of grouping, then, would rely on a bias mechanism.

Taking the perspective of evolutionary psychology, we might remark that the bias toward recognizing the next object, figure, shape, or structural element as "the same" would enable individuals or organisms to track down meaningful configurations in the shortest amount of time. If patterns are defined by the regularity of their structure, then the bias toward seeing one more of the same must be the simplest as well as the most strategic mechanism for pattern recognition. In one sense, this resonates perfectly with an observation in chapter 3, when it appeared that our cognitive system seems to be biased both toward thinking wishes to have come true and toward believing fears to have materialized. The bias was, generally speaking, toward the most meaningful interpretation—something rather than nothing, or a structure rather than chaos. Similarly, the proximity trap, with its bias toward what is close by, in time, space, or abstraction, is set up to favor the emergence of a figure from among the chaos or noise of the ground. The facilitation of pattern recognition in this way should easily benefit any species in terms of natural selection, providing it with a quick means to read a situation, address it, or "walk by in a curved line" as the poet Michael Palmer (1998, p. 41) put it:

The way the future uses up blood and light
and the individual marks are altered every day

until you reach the end of the row of trees
It has to be possible to imagine these

infinitely extended
and to walk by in a curved line

...

The excerpt is taken from "Fifth Symmetrical Poem" in *The Circular Gates*, a book of poetry with a definite geometric undercurrent that I cannot read without thinking, in a way circularly, about the emergence of gestalts in our thinking. Symmetry being one of the vassals of Prägnanz, I do suspect that thoughts coincident with that of the early twentieth-century Berlin school of experimental psychology were on the poet's mind, received straight from the muse if not from the papers by Wertheimer and Co. Here, "the way the future uses up blood and light," to my distorted ears at least, sounds like a lyric interpretation of BOLD data on prospective visual processing, recognizing a pattern from among the "individual marks" that "are altered every day//until you reach the end of the row of trees." What would be the use of the anticipatory representation? "It has to be possible to imagine these"—the trees—"infinitely extended/ and to walk by in a curved line." Vision takes a shortcut, and the seer will have the presence of mind and the necessary speed of thought to win the good, to flee the bad, or, in this particular case, to merely have a taste of endlessness, without ultimately getting trapped in infinite regress.

One of the principal tenets of Gestalt psychology was that much of the perceptual organization occurred autonomously, did not require any directed effort, and represented a true instance of dynamic self-organization. A particularly convincing demonstration to this effect comes from a study by Mattingley, Davis, and Driver (1997) with a patient who had suffered unilateral damage in parietal cortex and showed a cognitive deficit called "extinction," being able to detect single visual stimuli in either hemifield but ignoring one of two stimuli presented simultaneously in opposite hemifields (usually failing to report the stimulus in the hemifield opposite to the lesioned hemisphere). The deficit is thought to reflect a disruption of visual processing before selective "attentive stages of vision." The patient in question showed less severe extinction when the two stimuli in opposite hemifields formed a common surface. This was true even if the extraction of a surface required visual filling in, as with the well-known Kanizsa figures (which have a white figure, say a triangle or a square, emerging on a higher plane from out of the mouths of three or four little black Pac-Man-like critters). Without giving the matter any conscious thought, the patient was perfectly able to perceptually group the stimuli in accordance with Gestalt principles before selecting an object or a configuration of elements for the purpose of performing the detection task.

The case of the patient with parietal damage quite literally throws a plane over chaos, in a way that Deleuze and Guattari (1994, p. 197) would have recognized as characteristic of all human thought: "What defines thought in its three great forms—art, science, and philosophy—is always confronting chaos, laying out a plane, throwing a plane over chaos." For Deleuze and Guattari, this realization came toward the end of their journey, preparing the conclusion of *What Is Philosophy?* with the movement "From Chaos to the Brain." I like to think that it represents the (belated) discovery in continental philosophy of the fascinating world of biology. Deleuze and Guattari neglected to pun on the etymology of the word, but they would surely have appreciated the combination of the Greek words of *bios*, "life," and *logos*, "speech," to confirm that it is the prerogative of biology to talk about life, whatever its meaning.

If the interest in the brain came a bit late for Deleuze and Guattari, it must be granted that they were a few steps ahead of most neuroscientists today in thinking about the broader (philosophical, epistemological) implications of the physics of attractors and self-organization in chaotic environments. Particularly Gilles Deleuze shared with the early Gestalt psychologists a concrete interest in group theory and the works of mathematicians such as Bernhard Riemann and Henri Poincaré. For Deleuze and Guattari (1994), this approach held the promise of understanding the prime philosophical process of creating concepts or objects of thought—a process that, with Badiou (2007), we could call "positive" and poetic, positing things, as opposed to the subtractive or negative logic of mathematics and science, where we move systematically by rejecting false propositions. For contemporary neuroscience, it may be one of the most important challenges to try to translate the theories and formulations of chaos, self-organization, and attractor dynamics into viable and tractable hypotheses for empirical research. If Deleuze and Guattari (1994) were right about their intuition, neuroscientists will need to work in the same direction to gain a better understanding of how objects of thought are created in the brain.

A few bright minds in theoretical or mathematical neuroscience are already working in this direction (see Amari, Park, & Ozeki, 2006; Freeman, 2008; and Tsuda, 2001). Encouraging also is the emergence of a few empirical studies that explicitly incorporate analysis techniques borrowed from the mathematics of attractor dynamics (Denève, Duhamel, & Pouget, 2007; Ito, Nikolaev, & Van Leeuwen, 2007; Wills et al., 2005). These techniques allow researchers to move away from old and overly simplistic determinist models of neural functioning, toward a more accurate description of the stochastic nature of the underlying processes. Notably, the preferred data for interpretation in terms of dynamics have up to now been the rhythms or oscillations in electro- and magnetoencephalographic traces or in field potentials recorded from implanted microelectrodes, as well as the correlations in spike activity during so-called "multiple-unit recording" (recording simultaneously from several single neurons). Such data naturally look more "complex" or "chaotic," which is why old-school neuroscientists shy away from them, preferring simpler and more tightly controlled experiments that provide a manageable amount of data to confirm or disconfirm precise hypotheses. Nevertheless, brain waves and synchronized activity in local cell assemblies are very true phenomena that must somehow arise from what happens in single neurons. Researchers should do well, then, to try to examine systematically the relation between neural processing at different levels, say between spikes from single neurons and oscillations in the local field potential (as done very convincingly by Fries et al., 2001), or even beyond neurons, between hemodynamic responses and the activation of astrocytes, which may act as a support network for neurons (Schummers, Yu, & Sur, 2008).

One rather pedestrian way of moving forward with this systems-level approach in neuroscience would be to continue working with the tried and true simple and tightly controlled hypothetico-deductive ways, applied in Wurtz-like studies, and at the same time collect more data "on the side"—including field potentials, activities of multiple neurons, global brain waves, anything. The experiments are often very labor-intensive to begin with, and sometimes it would require relatively little effort to add a few channels, record something more. It is exactly this extra bit of information that could produce the analytic degrees of freedom to trace as yet unknown regularities, using theory-free or dimwitted algorithmic methods, among different levels or between different channels. I cannot help but see numerous wasted opportunities in most experiments, where researchers go to great lengths to train animals and carefully design perfectly sound experimental protocols but then harvest only a fraction of the data actually there for the taking. Here, the dictum of the French physicist, mathematician, and philosopher of science Pierre Duhem (1991) remains valid, suggesting that interpretation and observation are effectively inseparable in experimentation, or, as Hardcastle and Stewart (2003, p. 206) put it rather bleakly in relation to "the art of single cell recordings," neuroscientists collecting data today are operating in "the territory of educated guesswork."

Still, there is good hope that neuroscientists will yet fully appreciate the need to think about dynamics. As for the emergence of gestalts during perception, Rodriguez and colleagues (1999) have already been able to link the process of the creation of an object representation (a face) from an ambiguous figure with the appearance of synchronized brain waves in the gamma range (30 to 80 Hertz) across widely separated regions of the brain (see also Melloni et al., 2007). It looks as if the brain locks into a mode where the gravity of harmony creates objects of thought, twice a matter of "throwing a plane over chaos"—both in terms of the formal characteristics of the neural signals (now singing a song or dancing to a tune) *and* in terms of their representational properties (now indexing a figure in the world).

The work by Rodriguez et al. (1999) was situated well outside the mainstream of research in neuroscience when it surfaced, but the topic of neural interaction through neural synchronization appears to have finally gained ground (Kim et al., 2007; Womelsdorf et al., 2006, 2007). In a study by Womelsdorf and colleagues (2006), for instance, monkeys were required to report changes in one object while ignoring whatever happened to another object in the visual display. Intriguingly, the monkeys' reaction time in response to a visual change correlated with the degree of gamma-band synchronization among cortical area V4 neurons tuned to the target object. Moreover, this synchronization emerged well before the visual change occurred, suggesting that the gamma-band synchronization reflects an anticipatory process. Unfortunately, the authors did not apply behavioral analysis techniques to dissociate bias from sensitivity mechanisms. The paradigm certainly would appear to leave room for an effect of response bias in the change detection performance.

Could it be that the anticipatory gamma-band synchronization contributes to a neural mechanism of bias? Currently, the favored interpretation appears to go in the other direction, proposing that the gamma-band oscillation in fact contributes to the multiplicative scaling of tuning curves (Kim et al., 2007; Zeitler, Fries, & Gielen, 2008). Here, the assumption is that the synchronization in the gamma range improves the synaptic gain for spikes in response to the target stimulus. It is a stimulating discussion, in any case, and I do hope researchers will employ methods such as those from signal detection theory or the LATER model to explicitly consider the possibilities of bias versus sensitivity mechanisms (or bias *plus* sensitivity mechanisms) instead of making the usual, premature leap from a behavioral paradigm of "selective attention" to multiplicative scaling or improved signal-to-noise processing.

With the discussion of gamma-band oscillation, we have moved from the self-organization of object representations in perception to the domain of attentive, conscious, or deliberate sensory information processing for the purpose of performing a task. Whether or not the gamma-band oscillation will indeed turn out to be related to the quality of sensory information processing, enhancing signal-to-noise ratios and scaling tuning curves multiplicatively, a second important question is how the oscillation comes about—how is it controlled? From the bottom up, exogenously, stimulus driven, or rather voluntarily, endogenously, depending on the beliefs and expectations of the observer? The question was high on researchers' agenda in the 1980s and 1990s (Jonides, 1981; Lauwereyns, 1998; Theeuwes, 1994) but has since drifted to the background. I think we may expect to see a revival of interest in this question once researchers start focusing more sharply on the transitions from one brain state to another, or the temporal dynamics of the emergence of selective information processing.

A particularly promising phenomenon in this respect is that of "pop-out" in visual search (see Smith, Kelly, & Lee, 2007, for relevant data from V1 neurons), when a singleton target (say a red element among a homogeneous set of green distractors) produces flat search slopes, independent of the number of distractors in the display (Treisman & Gelade, 1980). According to Treisman's feature-integration theory (see also chapter 1), the flat search slopes would be indicative of parallel visual processing, requiring no intervention from "attention" (thought of as a cognitive mechanism whose function it is to glue together any features that are found in the *attended* spatial location). With the concepts of Gestalt psychology, however, we might reinterpret the preattentive search in terms of gestalts emerging from autonomous, self-organizing processes.

The principles of Prägnanz and the workings of the proximity trap would generate a gestalt from processing the set of similar items (in my example, the group of green distractors). The underlying mechanism, I propose, would be that of a bias presuming the entire field to consist of more of the same, gaining force with every new item that does indeed answer to that call. Literally in a matter of tens of milliseconds, the visual cortex would have detected the pattern (or "context") of green items. Yet, then there is this item, this lone red one that refuses to listen to the call. It becomes the instigator of a symmetry-breaking event, to use the language of attractor dynamics—a bifurcation happens; a new figure, a singularity, breaks out ("pops out") of the pattern of green. Now this singularity takes possession of the mind, "in clear and vivid form" (James, 1950, p. 403), even though the feature integrator was not asked to integrate any features. It is all a matter of self-organization, fueled by contextual bias, and counter-self-organization, when the evidence overpowers bias to the point of tipping the scales and becoming the new focal point of attraction.

Only in more noisy or ambiguous visual environments may we expect to find a dominant role of top–down or voluntary processes in the definition of objects of thought. There must be something of a dynamic range for perception—a fuzzy area where the physical stimulation is too weak or too complex to impose a configuration. Within that flexible range, the observer would be able to set the threshold or spatial resolution for grouping by proximity (Logan, 1996), or to regroup or connect the dots in more idiosyncratic ways, inventing "new constellations" (to quote a title by the Flemish poet Peter Verhelst, 2008). In this last case, we decidedly work against the lure of the proximity trap, creating objects so creative that our friends and colleagues, or even the whole of contemporary humanity, may fail to understand exactly what kind of thing it is we think we see. Or if successful, ours may prove to have been a new way of making history. In the words of Michel Foucault (1994, p. 131), discussing the pioneering work in biological taxonomy by Johannes Jonstonus (John or Jan Jonston), the seventeenth-century Polish scholar who clearly wished to distinguish fact from fiction in natural history:

What came surreptitiously into being between the age of the theatre and that of the catalogue was not the desire for knowledge, but a new way of connecting things both to the eye and to discourse. A new way of making history.

Throughout the renaissance, Foucault noted, the strangeness of animals had been a spectacle, but ever since Jonston the awe for the mystery of the unknown, and its relative passiveness and receptiveness in being struck, had had to move aside for the more enterprising attitude of scientists, active in experimentation, creating new gestalts, grouping things in different ways, even if the cycle from theory to data and back often looked like a paradoxical algorithm that only served to generate more questions and bring about the eternal return of the mystery of the unknown.

The Trouble with Eternal Return

When the Gestalt psychologists formulated the principle of Prägnanz, the law of proximity seemed to pertain specifically to *different* items that, by virtue of being in each other's neighborhood, could be understood to belong in one and the same perceptual group, spread out spatially over a visual plane, temporally across time, and so forth. However, what should we say about items that reappear after a short interval at the same (or a nearby) position in space? When do we have two items that we can recognize as members of the same group versus one item that we observe translating across time and space? Do we have a return of the one or a more abstract return of another *like* the first?

The issue comes back in different guises, such as the question of object constancy or the topic of neural adaptation. The core problem is that of correctly identifying changes in our sensory environment—changes due to the movement of objects, the shifting position of the moving observer, or both. The challenge is complicated by several factors, but it is perhaps most pervasively shaped by the degrees of freedom (or the margin of error) in neural coding. Here, the definition of sameness for any bit of information is given by the resolution of its neural representation. If we can say, for instance, that the spiking rate of a cortical area V4 neuron represents, conveys, or otherwise correlates with the physical presence of a particular color, then we can examine the neuron's tuning curve to obtain a computational measure of the fuzziness of this code. However, the tuning curve must have temporal properties as well. The neuron will show a peak of activity in response to the onset of its "coded" stimulus within a certain amount of time—say, three hundred milliseconds—and then the neural activity will dissipate in a process that presumably corresponds with adaptation. This temporal process, the adaptation, brings with it a different kind of tuning, or another set of degrees of freedom, with implications for when the neuron would be ready to signal new information, a change, or the occurrence of a different item.

The faster the adaptation, the better (or the more dynamically) the neural circuit will be able to respond to what happens next in the sensory environment. The process itself would be adaptive—an adaptive form of adaptation—in the sense that neurons can change their tuning dynamically to match changes in the sensory input (Dragoi et al., 2002; Sharpee et al., 2006). An early demonstration of this phenomenon was the apparent predictive remapping of visual space, when the receptive fields of visually sensitive lateral intraparietal neurons shifted as a function of an intended eye movement even before the monkey initiated the eye movement (Duhamel, Colby, & Goldberg, 1992). The neuron was one step ahead, already adjusting to the future of what would follow after the eye movement. The predictive remapping extends to nonspatial visual features (Melcher, 2007) and so would seem to be a perfect mechanism (anticipatory and biased) to facilitate the tracking of objects across eye movements, already thinking to "see" objects where we should expect them to be after the eye movement—a mechanism to enhance the sense that the same thing remains the same even if it now excites an entirely different set of photoreceptors.

If the adaptability of adaptation is tailored toward maximizing our ability to compute object constancy, then one corollary of this bias in favor of the "same" would be that the proximity trap across time in fact generates a peculiar kind of perceptual grouping, in which different items are grouped together not as a field of similar items but as a sequence of different views of the very same item. Where the group or the whole used to be of a different order than the individual items, now the group is in fact interchangeable with the individual in the case of repetition of items across short time intervals. Indeed, if the time between stimulus presentations is short enough, we will not even be able to notice that there are multiple presentations—the images will fuse above a critical flicker frequency (a phenomenon known at least since the 1830s, when the Belgian physicist Joseph Plateau invented his *phenakistoscope* to cheat the visual system into believing that drawn pictures were actually animated and moving).

When the interval between stimulus presentations is on the order of seconds or hundreds of milliseconds, we will be aware, at least in visual or auditory processing, of a gap, separating two distinct events. Yet, even with the noticeable delay, the processing of the second stimulus will be affected by the "prime," usually with an increase in response speed when the second stimulus matches the first as compared to when the two stimuli show little resemblance to one another (Entus & Bindra, 1970; Lauwereyns et al., 2006), though the opposite phenomenon—*negative* priming—may occur when the subject is required to ignore the first stimulus, in which case it appears as if the initial suppression of a perceptual representation carries over to subsequent perceptual processing (Tipper, 1985).

As for the "default" case of positive repetition priming, a particularly spiky debate in cognitive psychology in the early 1980s suggested that the performance enhancement could be due to improved sensitivity (Proctor, 1981), bias (Krueger & Shapiro, 1981), or both (Proctor & Rao, 1983). Overall, the behavioral data appeared to favor the second Proctor publication, which did manage to put a stop to the ping-pong in *Psychological Review*. What do the neural data say? Intuitively, the proactive interference on the second stimulus presentation would seem to pertain to some kind of residual activity in the neural correlate of a perceptual representation (see figure 5.3, panel b). In the case of color repetition, we can expect cortical area V4 neurons that are tuned to the presented color to show some activity lingering for a while after the presentation of the first stimulus. When the second stimulus arrives, the processing would start from an elevated baseline, exactly as I have discussed a number of times already with the LATER model.

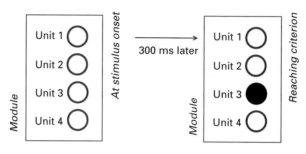

a

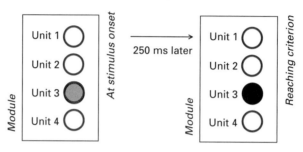

b

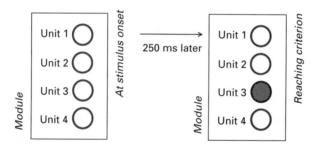

c

Figure 5.3
Neural correlates of repetition priming, hypothesized and observed. (a) Activity in a visual feature module (e.g., cortical area V4 for color processing) upon the first presentation of a stimulus. The (simplified) module consists of four units (or types of neurons), tuned to different features within the same stimulus dimension. At the time of stimulus onset, all units are firing at their baseline rate. After a certain time (here, marked as three hundred milliseconds) the unit tuned to the actual stimulus feature reaches its activity peak. (b) Hypothesized correlate of repetition priming. At the second occurrence of the same stimulus, the relevant unit would still be slightly more active than other units due to residual activity in response to the previous stimulus presentation. As a result, the unit would start from a higher activity level and reach its peak activity sooner than with the first stimulus presentation. (c) Empirical data show a different pattern, with no differences in the spiking rate at the second onset of the same stimulus and with an earlier but also lower activity peak as compared to that in response to the first stimulus presentation.

The relevant neurons would reach the threshold for the perception of the presented color faster than without the prime, explaining the reduction in response time as an effect of bias. Here, the putative residual activity is taken be an "anticipatory" process, not necessarily prospective but clearly antecedent to the onset of the second stimulus. So much says the hypothesis. The data, however, refuse to listen to intuition (see figure 5.3, panel c). In single-unit recordings, the monkey's behavior may show evidence of repetition priming while cortical visual neurons that are tuned to the presented stimulus show no trace of any residual activity in their spiking rate (McMahon & Olson, 2007).

At the onset of the second stimulus, all neurons of a visual module are at their base rate of activity, and perceptual processing would seem to start from a blank slate. In fact, the effect of priming is primarily visible as a reduction of the peak activity of neurons tuned to the presented stimulus. This is surprising, given that we are used to seeing a negative correlation between the strength of a neural response to a stimulus and the speed of a behavioral response to that stimulus. Instead, in this case a weakened neural signal is associated with an improved behavioral response. The weakened neural signal in fact implies a reduced signal-to-noise ratio, so I think the best explanation is still in terms of a bias mechanism.

If we assume that the strength of the neural signal after stimulus onset reflects the amount of "push" or energy required to reach a decision threshold (as suggested by Hanes & Schall, 1996), then we can interpret the reduced peak activity to reflect a change in criterion, or a lowered threshold. Put differently, the reduction in the amplitude gives a metric of the extent to which the distance between the starting point of the decision process and the threshold value was shortened. The cortical visual activity, correlating with the sensory information processing, might have had to work less hard for the second stimulus presentation thanks to a change in the baseline activity elsewhere. If so, we should look to neurons one level up in the hierarchy of decision making—neurons that receive input from the cortical visual neurons and gate the information on toward motor structures, neurons in the prefrontal cortex (Kim & Shadlen, 1999; Sakagami & Tsutsui, 1999). It could be that prefrontal neurons start from a higher base rate of activity after the first stimulus presentation, so that they now require less evidence or fewer spikes from cortical visual neurons that signaled something similar a moment ago.

Is it reasonable to assume that prefrontal neurons show a heightened baseline firing rate due to residual activity? Prefrontal neurons certainly can show anticipatory processing in advance of a task-relevant stimulus (Sakagami & Niki, 1994) as well as sustained activity following the presentation of a task-relevant stimulus (Funahashi, Bruce, & Goldman-Rakic, 1989). To my knowledge, the relevant data remain to be collected. Interestingly, moving from prefrontal cortex toward motor structures, we do already have evidence of elevated anticipatory processing as a function of repetition priming. Dorris, Paré, and Munoz (2000) found that the preparatory activity in superior colliculus neurons prior to a target stimulus was higher when the previous stimulus had appeared in the same position as compared to the alternative position in the opposite hemifield.

Piecing everything together, then, it does look feasible to construct a coherent explanation of repetition priming as a result of bias, with a proximity trap in the temporal domain. Data from fMRI experiments further underscore the notion that repetition priming produces the counterintuitive combination of improved behavioral response with reduced neural signaling (Grill-Spector, Henson, & Martin, 2006; Krekelberg, Boynton, & Van Wezel, 2006). Nevertheless, the bias interpretation may not tell the complete story. It is certainly possible that a synergistic process is at play in repetition priming as well, involving some kind of intracellular or synaptic mechanism that does not depend on continued spiking activity (see Mongillo, Barak, & Tsodyks, 2008, for a convincing proposal). Mechanisms of this kind would work selectively in a way that might in fact sharpen the tuning curve of neurons and, thus, ultimately imply an increased signal-to-noise ratio, specifically in comparison with stimuli that are in the middle range on the tuning curve—not the "coded" stimulus or the "other" stimulus but "similar" stimuli from somewhere in the transition areas between the extremes of the bell curve.

Perhaps a little numerical exercise can illustrate my point. Let us consider three points on the tuning curve (harking back to the labels introduced in chapter 1): (1) the apex, for the "coded" stimulus; (2) the middle, for a stimulus that is relatively similar to the "coded" stimulus; and (3) the bottom, for the "other" stimulus, maximally dissimilar to the coded stimulus. Upon first presentation of a stimulus, the neural responses might be eight spikes at the apex, four spikes in the middle, and two spikes at the bottom of the tuning curve. Now, upon second presentation, the responses could be six spikes at the apex, two spikes in the middle, and two spikes at the bottom of the tuning curve.

In this carefully contrived example, we can see a hybrid of bias and sensitivity at work. Comparing the neural responses for the "coded" stimulus to those for the "other" stimulus, we go from an eight-to-two ratio upon first presentation to a six-to-two ratio upon second presentation, or a reduced signal-to-noise ratio, calling for an explanation in terms of bias. However, comparing the neural responses for the "coded" stimulus to those for stimuli from the middle of the tuning curve, we reach a different conclusion. We go from an eight-to-four ratio to an actually improved ratio of six to two. The reduction of the peak activity would still be best explained as a consequence of bias, but the additional changes in the tuning curve are described most appropriately as a sensitivity effect, not scaling the curve multiplicatively but in fact reshaping it. Proctor and Rao (1983) might have been right after all.

To be sure, my example is entirely conceptual. Yet, the question of how tuning curves change as a function of repetition priming is one that can and should be addressed empirically, as the changes are likely to depend on the length of the temporal interval between stimuli as well as on the type of neuron under consideration (see Verhoef et al., 2008, for the perfect case in point). With this observation, we return to the basic issue, the trouble with eternal return, for what is it exactly that returns, or when are two stimuli sufficiently similar for the one to serve as a prime for the other in a way that allows us to say that we see an effect of repetition priming?

And what is it exactly that we decide in that case? Are we simply better at seeing the second stimulus due to the accidental similarity with something in the past, or do we see a link, make a connection between the two events, saying something about their identity—the identity of the one and the identity of the other as well as the identity of the one *and* the other as the very same thing viewed twice. If the questions start sounding slightly philosophical, it might be because that is what they are. As it happens, they have been explored thoroughly by Gilles Deleuze (1994) in *Difference and Repetition*. In his take on the Nietzschean concept of eternal return, even philosophers would rather bypass the tricky question of identity. Deleuze (p. 241) writes:

When we say that the eternal return is not the return of the Same, or of the Similar or the Equal, we mean that it does not presuppose any identity. On the contrary, it is said of a world *without identity*, without resemblance or equality.

What returns, then, is not something identified. Rather, eternal return would be something that happens in the imagination first—the moment "a thing" is thought to return is the moment that defines the thing and retrospectively induces sameness. It is the noticing of a connection, a common thread, or a likeness that gives structure to the object or gestalt, as a corollary to the very act of grouping, which is why perceptual grouping can rightfully be called an act of *organization*. This is simply saying that the definition of an object as a collection (or grouping) of features—the semantic properties or content—is not conceivable without the simultaneous definition of the object's form—its geometric properties, its coordinates in time and space.

A Certain Surplus

Noticing connections, common threads, or likenesses that give structure to objects... it brings to mind the original meaning of religion, in Latin, *religio*, derived from *religare*, to tie fast, to bind together. Combined with the etymology of poetry, it would seem the prerogative of the seer, the prophet with the vatic powers to transcend the here and now, to group and organize things perceptually in new and meaningful ways. Less grand and more to the point, we might refer to a certain surplus as one of the fundamental aspects of attractor dynamics and nonlinear self-organization from chaos—a certain surplus in the emergence of something more than the sum of its parts, an added quality of a different category. The gestalt that arose from the connections among individual elements belongs to a different plane and becomes visible from a vantage point where there was absolutely nothing to see before, nothing that could predict the appearance of an entirely new entity on the scene. Deleuze (1994, p. 70) discussed this curious phenomenon of a certain surplus as a paradox at the heart of repetition:

> Does not the paradox of repetition lie in the fact that one can speak of repetition only by virtue of the change or difference that it introduces into the mind which contemplates it? By virtue of a difference that the mind *draws from* repetition?

Repetition causes a difference in the mind, Deleuze argues. The drawing of difference from repetition is "the role of the imagination" (p. 76), or we could even conclude that the repetition in itself is "in essence imaginary" (p. 76). Physically speaking, we have nothing but a set of statistical regularities among elements in time and space, but in our mind, these regularities are read as structures that repeat. And so, on second thought, we do note a difference, even physically speaking, in what happens to our neural circuits when the repetition of one thing or another reaches the status of concept, idea, or something mental we can recall and communicate.

Deleuze (2004) further hints at something of a mathematical law for the imagination, suggesting that the perceived repetition "weights the expectation in the imagination in proportion to the number of distinct similar cases observed and recalled" (p. 71). The repetition causes a difference in the mind, or a redistribution of weights in neural circuits, in such a way as to create a bias in the expectation of what comes next, of how features are associated with other features in objects. Thus, we move from the proximity trap, where things are grouped by their similarity or the apparent repetition, back to the familiarity fallacy, where expectations are formed on the basis of the frequency (or the amount of repetition) of co-occurrence among events.

In both the familiarity fallacy and the proximity trap, we have seen bias at work as a pervasive property of information processing that provides us with the power to imagine and thereby tailor the future and the ability to create objects, perceptual groups, things of the mind. Bias operates at the heart of gestalt formation and so must deserve some credit for the nonlinear emergence of a whole that is more than the sum of its parts. Yet, the transition from nothing to something baffles us, takes us unaware every time. The process remains mysterious and magic, and even if we think we know the formula and are able to reconstruct how things that occurred did, in fact, manage to occur, no one seems to be able to tell us exactly how to make it work in the other direction, prospectively. We are not able to control it or to apply it ad lib. Suddenly the awesome glow is there, as with great poetry:

In each successful instance, a certain surplus of meaning will occur, a dimension of semantic inexhaustibility beyond the strictly communicative function, that will allow the poem to return, will allow it to converse with an *other*, an unknown addressee, will allow it to flow and submit to alternative readings, alteration, othering. Here is it that form's mind insists, beyond whatever practical or aesthetic intent, on the *more*. (Palmer, 2008, p. 79)

The unpredictable *beyond* astonishes us with its apparent infinity, but when does it work that way? "In each successful instance," Michael Palmer (2008, p. 79) writes in *Active Boundaries*, a collection of essays. The answer refuses to be an answer. We can only guess how much error in the trying there has to be for every poem that succeeds. That the being of the gestalt, or the nature of the certain surplus, escapes our grasp may be cruelly (or deliciously) ironically the principal condition for its aesthetic appeal. Something of the sort ventured Walter Benjamin (1968, p. 199), one of the most celebrated literary theorists of the twentieth century. The proposal vindicates the autonomy of attractor dynamics and self-organization in perceptual grouping, as according to the first wave of Gestalt psychologists.

In the meantime, another paradox has worked its way into the discourse on perceptual grouping—a paradox that we can aptly summarize with the wonderful aphorism and motto of minimalism, "Less is more" (most often attributed to Ludwig Mies van der Rohe, a pioneer of modern architecture). Indeed, from the reduction of many elements to one gestalt, we gain the immensity of something unique, a special object, a unit of thought, inviting many returns, and much conversation, replete with alternative readings, alteration, othering. This paradoxical effect of selection and reduction is fundamental enough to the whole enterprise of decision making to warrant a chapter of its own, so let us make it the next.

6 Less Is More

The Soul selects her own Society—
Then—shuts the Door—
To her divine Majority—
Present no more—

Unmoved—she notes the Chariots—pausing—
At her low Gate—
Unmoved—an Emperor be kneeling
Upon her Mat—

I've known her—from an ample nation—
Choose One—
Then—close the Valves of her attention—
Like Stone—

This is poem number 303 by Emily Dickinson (1961, p. 143), the most private and introspective poet known in the English language, who, in total obscurity in the middle of the nineteenth century, somehow managed to explore the riches of the human mind in a way that still fascinates and challenges readers today. ("Still" is not the right word—the interest in Dickinson's poetry is probably stronger today than it has ever been.) She worked in the constrained space of a few stanzas, with the dash as her principal instrument of punctuation, setting the speed of thought, and visualizing the gaps and leaps in her elliptical composition. No poetry that I can think of better illustrates the power of concentration and the paradoxical truth of "Less is more" in the selection of just a few images and a handful of words to reflect the movements that constitute our thinking, our experience of a particular moment.

Number 303 does this recursively on the topic of choosing or shutting the door, on the relationship between attention and being moved, here phrased in the negative, remaining unmoved by what the Soul has locked out. Dickinson uses words and concepts that look anachronistic, several decades ahead of their time ("The Soul selects her own Society" and "the Valves of her attention," closing "Like Stone"). The poem in my mind echoes later texts by William James (1950) and Aldous Huxley (2004) on the inhibitory mechanisms that are at play when an object or train of thought takes center stage in our conscious experience, shutting everything else out. Of course, it is in the nature of the beast that I am to notice that these mechanisms imply a bias, with preconceived labeling and additive (in this case subtractive) scaling—for except for the happy few that make the cut, the whole wide world has to endure suppression, facing a heightened threshold, a longer journey in the LATER model, or a more stringent criterion within the framework of signal detection theory.

There is good reason to apply this negative form of a bias against the masses of information. For one thing, the need for selection necessarily derives from the obvious truth that all living creatures, from fruit flies to humans, find themselves moving within a narrowly defined field, with limited degrees of freedom—a constrained "answer space," in which the number of possible responses is largely defined by very real and physical factors, including the dimensions of the body and its surround. We only have two hands and one pair of eyes, and with that we can try to do what we can do.

A Bias toward Bias

Already in 1949 Donald Hebb had noted that "[t]he lack of a rationale for nonsensory influences on behavior that seemed to exist in 1920 certainly exists no more" (Hebb, 2002, p. 7). To reject the assumption of sensory control had, in fact, become something of a point of departure in the development of a neurophysiological theory of the organization of behavior. "Man or animal is continuously responding to some events in the environment, and not to others that could be responded to (or "noticed") just as well," wrote Hebb (2002, p. 4). The statement seems so obviously correct today that it is hard to fathom how anyone could have failed to incorporate it in the theorizing on mind and brain, unless we try to imagine a hardheaded scientist willfully creating a theoretical caricature, mercilessly applying the maxim of "Less is more" in the design of flowcharts between input and output, scrupulously avoiding anything that remotely reeks of anthropomorphism or animism in the gating of signals.

Donald Hebb (2002) was just one of several notable voices that emerged in the 1940s and 1950s claiming a central role for selection in information processing. Other luminaries included James Gibson (1950) and Donald Broadbent (1961). Broadbent advocated a theory of early selection, arguing that only the most cursory analysis occurred before an item or object was selected for further information processing, whereas Deutsch and Deutsch (1963) provided data to show that selection sometimes took place after items had been processed all the way up to identification—the issue apparently got sorted in the 1990s when Lavie and Tsal (1994) offered that perceptual load determined the locus of selection (see Van der Heijden, 1992, for a similar suggestion; Desimone & Duncan, 1995, provided a popular neural analogue with the "biased competition" model—using the term *bias* in a slightly different way than I do here).

Meanwhile, metaphors of selection abounded, with the spotlight (LaBerge, 1983), the zoom lens (Eriksen & St. James, 1986), and the bottleneck (Pashler, 1984), and researchers started discussing selection at the level of motor processing (Allport, 1989; Rizzolatti et al., 1987). Most of these theories, models, descriptions, and speculations appeared to incorporate aspects of both facilitation and inhibition—facilitation of whatever information is selected, illuminated, in focus, or squeezed through the narrow cylinder of the bottle's neck, and inhibition of what fails to make the cut, remains in the dark, blurry, or stuck at the same side of the bottle, be it in or out. Facilitation invariably implied the operation of sensitivity mechanisms, improving the quality of information processing, with brighter, sharper, and more elaborate representations of the selected information. Inhibition seemed to come in the guise of a negative bias, bringing with it an anticipatory form of subtractive scaling (see Sylvester et al., 2008, for a demonstration in BOLD responses).

Early-selection proposals such as those by Broadbent (1961) and Posner (1980) leaned particularly heavily on the improved information processing for the selected information. Late-selection accounts seemed intuitively more compatible with mechanisms relying on noise reduction or suppression. However, even if the parameters and conditions of selection in information processing now stood squarely in the foreground of research programs in psychology and neuroscience, the very raison d'être of selection proved to be a more elusive topic. Usually, the first idea that comes to researchers' minds is that selection is an adaptation to limitations in resources or capacity—a good idea only if the degrees of freedom are attributed to the right level of processing (Pashler, 1994; Lucas & Lauwereyns, 2007). Taking the view of resource limitations to the extreme, it implies that the more resources we have, the less we would depend on selection; with a bigger brain we would not prioritize as much, show less of a bias, be more neutral, and care less. I do not think this tells the whole story. I believe selection would still be at least as selective and as affected with a genius brain as with a below-average one. The difference, I think, is rather to be situated in the process of selection itself, the algorithms of selection, and the extent of exploration before selection.

I prefer to think of selection rather as a mechanism that provides a gain—that gives the "more" in "Less is more"—than as an adaptation to the excess of information available in the environment, that is, a mechanism that limits the losses by losing as much as possible as early as possible. The function of selection is not merely to keep things, but to work on them, with them, creating a future, a response, an interaction. The process of selection itself may already serve as a preparation for what comes next. That is, selection should be viewed more dynamically, not as a divide between a before and an after but as a stage of processing with its own properties and consequences. Applying some analogical thinking, I find myself remembering the remarkable string of world records (plus one Olympic record) that Michael Phelps and his relay partners swam at the Summer Olympics in Beijing in 2008, winning eight gold medals. It must have been the extraordinary circumstances, the big prizes on offer, the powerful rivals (László Cseh, Milorad Čavič, and many more), that conspired to produce this unprecedented achievement.

The selection of "the number one" created the condition for truly outstanding performance. The idea that some form of competition among candidates or alternatives can modulate the speed and accuracy of behavior has been incorporated in a variety of decision-making models, whether purely behaviorally or more neurophysiologically oriented. Particularly the stop-signal paradigm has proved to be a fruitful platform for the generation of "horse race models" (Band, Van der Molen, & Logan, 2003). Here, the simple assumption that several stochastic and independent processes (horses or competitors) run in parallel would already suffice to produce a positive correlation between the number of competitors and the performance level of the winning competitor. The more horses, the faster will be the winning time.

Intuitively, we might think this proposal to be incomplete. Any candidate is likely to change his or her behavior in the face of stiff competition—the energy levels go up, more effort is invested, and so forth. Computationally, we could state that the assumption of independence among competing processes is likely to be incorrect, and indeed, neurophysiological investigation suggests that a neural network of interacting units provides a better fit with behavioral and single-neuron data than a model that assumes independence among stochastic processes (Boucher et al., 2007). The pressure from competition does alter the course of decision processes. When the stakes are raised, the information processing intensifies, and different alternatives might undergo a fuller screening than if the situation was more noncommittal, things did not matter all that much, or there was essentially nothing to be gained or lost.

Thus, I find myself advocating, for once, an overlooked potential for effects of sensitivity where models of selection have all too often equated losing with being suppressed or getting the most casual once-over. Even if the winner of the competition gets to take all—be it "attentional priority" (Bisley and Goldberg, 2003), a place in working memory, or some other privileged opportunity to be the object of extensive cognitive processing—the loser may have exhibited improved performance, or, phrased passively, the loser may have been processed more extensively. Empirical support for this line of thinking is scarce, I will readily admit, but not nonexistent. An intriguing study by Williams, Henderson, and Zacks (2005) recorded, in a token-discrimination task, how much detail subjects remembered incidentally of targets and distractors that had been presented in an earlier visual search task. As it turned out, visual details of distractors were remembered about as clearly as those of targets, but only if the distractors were in some way "related" to the search target, that is, if they had generated stiff enough competition.

Combined with the more commonly accepted notion of improved information processing for the selected object or item, then, we have two good reasons to employ a systems bias for bias, or a general preference for the application of preferences even in situations that do not structurally require picking one alternative and removing the others. The two good reasons come down to two ways of interpreting the maxim of "Less is more." There is the "Less is more" of the glow and the special treatment that befall the selected item, now being looked at, scrutinized, revered, and analyzed to bits. And there is the "Less is more" in the slightly more convoluted sense that claims the condition of competition to be conducive to an overall improvement of performance, with better winners as well as better losers in a spirit that may not be exactly altruistic, gracious, or Olympian but certainly generates world records.

Happy Accidents in the Evolution of Ideas

The systems bias for bias, with its capitalization on the *more* to be gained from the *less*, is, of course, only as good as its definition of goodness. In applying a bias, and instituting a competition, we pull forward a target, a template, a specified set of features considered relevant. While this prospective representation of a winner of the competition can improve the speed and accuracy of relevant responses, it does inevitably also limit the degrees of freedom for what is imaginable as a relevant response. The competition sets the rules by which the players must play in the hopes of winning the big prize ("all"). The downside of this preconceived domain of relevance is that it can literally make the observer blind for anything that does not fit the mold of relevance.

The point was already made more or less explicitly with the reference in chapter 5 to the proximity trap in the phenomenon of "change blindness"

(Simons & Rensink, 2005). Some forms of change blindness could be seen as collateral damage to the systems bias for bias, particularly when the definition of relevant information for processing creates a blind spot for other important information of a more unexpected variety, somehow not matching with the current definition of relevance. Mack and Rock (1998) investigated the operation of this kind of inflexible and inadequate selectivity with an impressive sequence of behavioral experiments under the label of "inattentional blindness" (see also Koivisto & Revonsuo, 2007, for an unusually dramatic demonstration). The common thread in these demonstrations appears to be that the detection of salient but irrelevant stimuli depends on their degree of similarity with any item from the predefined set of relevant information.

Figure 6.1 shows a perfect example of inattentional blindness to which I should plead guilty. Mesmerized as I was by the performance of number 81, enjoying the "Less is more" in contemplation, I (the little figure inside the

Figure 6.1
A clear instance of inattentional blindness during the annual relay race of the U.S. National Institutes of Health in Bethesda, Maryland, in 2002. Number 81 (Okihide Hikosaka) displays great physical fitness, while the distant figure in the background (indicated with a white quadrangle) is so fixated on the race that he forgets his (or I forget my) duty as a traffic controller. The white van should have stayed behind the traffic cone (cf. next to the right foot of the person struck by inattentional blindness).

quadrangle) found myself completely oblivious to my main task of controlling traffic—I had literally not seen the rather sizeable white van drive beyond the traffic cone where it was supposed to stop. The van snuck past me, ready to go onto the track of the relay race that I was resolved to protect from exactly this type of intruder. In this case, my systems bias for bias did not manage to keep up with the correct order of priority, switching the actually relevant set ("select any information relating to the task of controlling traffic") for a set that naturally stimulated my curiosity but should have remained secondary ("select any information relating to runners from the Laboratory of Sensorimotor Research")—perhaps my systems bias for bias fell prey to the familiarity fallacy. Luckily, the driver of the van realized, completely of his or her own accord, that it was a good idea not to drive into the crowd of runners.

Inattentional blindness set aside, the systems bias for bias in the application of "Less is more" generally offers more (tautologically more) benefits than costs. This is true for the heuristics that Daniel Kahneman, Amos Tversky, and their associates identified in the way people make judgments and decisions (see Kahneman, Slovic, & Tversky, 1982, for what is still the best point of departure in this domain). Simple rules of thumb work best to scan situations quickly, to find solutions without too much effort, and to move ahead with a course of action—often a good idea, because even if the chosen solution is not the best solution, it might still be a good enough solution, better than wallowing in a state of indecision (bringing to mind again Herbert Simon's (1956, p. 129) dictum that "organisms adapt well enough to 'satisfice'; they do not, in general, 'optimize' ").

The wisdom of "Less is more" has certainly been appreciated in philosophy as well, perhaps most famously by Ludwig Wittgenstein, both the early Wittgenstein of *Tractatus Logico-Philosophicus* (2001), working toward a geometric system of facts and indisputable truths, and the late Wittgenstein of *Philosophical Investigations* (2003), having abandoned geometry as a model for thought and working more organically, associatively and metonymically, with observations and carefully phrased insights. The human mind's penchant for simplicity is probably there for good reason, as W. V. Quine (1966, p. 209) pointed out:

Biases in our conceptual schemes may have great utility in the systematizing of science, and therewith high survival value, despite humble origins in the random workings of unreason—just as chance mutations in the chromosome may launch a sturdy and efficient new race. Natural selection through the ages tends to favor the happy accidents at the expense of the unpropitious ones, in the evolution of ideas as in the evolution of living species.

Biases work beautifully to organize information as we have seen on the topic of the proximity trap. Fixations, obsessions, monomaniac interests all come with their thematic implications, offering the recurrence of a motif, the repetition of a limited set of ideas, the ever-present melodic phrase in a musical composition. Such is the power of idiosyncratic *topoi*, or the lure of semantic attractors, that it can organize the most linear of arts, the novel, into a collection of perambulations that provides no easy story yet manages to be immensely readable—I am thinking of *Austerlitz* (2002) by W. G. Sebald, for instance, or *Underground* (1997) by Don DeLillo.

In the present operation of a systems bias for bias, I might add, the recurring motif of *The Anatomy of Bias* goes one better still, with another layer of recursion, the writer writing with a steady focus on the topic of bias—and, come to think of it, in this very sentence I am probably pushing it a bit, writing about the writer writing on the topic of bias. Before the wild looping becomes too much, I will do well to quickly shift the focus of the current sentence back to W. V. Quine (1966, p. 210) and the usage of simplicity as an implicit but pervasive principle of thought: "[H]ow do we decide, apropos of the real world, what things there are? Ultimately, I think, by considerations of simplicity plus a pragmatic guess as to how the overall system will continue to work in connection with experience." The default attitude, the one that gains time, affords action, and gives the best return, is to take things for what they appear to be, to go with the easiest interpretation, to assume that if something walks like a duck and talks like a duck, it might actually *be* a duck. Yet, W. V. Quine (1966) correctly pointed out that there is no privileged relation between simplicity and truth, despite the frequent reference in science to Occam's razor. No sound logic or theory allows us to claim that the simpler explanation must of necessity be the correct one.

There exists, on the other hand, a privileged relation between simplicity and our capacity for thought, the amount of information we can keep "online," the number of things we can keep track of in the hopes of responding adequately to them. There is probably also a privileged relation between simplicity and our capacity for aesthetic appreciation, under the right circumstances at least—no doubt scholars trained in evolutionary psychology will be able to provide accounts of how it would benefit this or that species to develop a liking for this or that kind of symmetry, wholeness, or smooth texture (see Rhodes, 2006, for an overview with respect to humans' sensitivity to facial beauty). Ultimately, though, the wisdom of "Less is more" appears to pertain first and foremost to the pragmatic domain. On a good day, the systems bias for bias allows us to exploit in the most strategic way the variety and competition among alternative objects or trains of thought.

Monkey Stroop

On a bad day, the systems bias for bias gets to deal with a kind of competition that it had not bargained for. The information that it wishes to pick out from among the mass of signals impinging on the senses turns out to lend itself less easily to digestion than expected due to the presence of something else, some competing information that really should be ignored, stopped, held back, but somehow manages to leak through. The "gold standard" of demonstrations to this effect is the Stroop test (MacLeod, 1991a, 1992), developed by J. Ridley Stroop in the 1930s (Stroop, 1992). The core idea is, to speak with the aesthetic side of "Less is more," beautifully simple. Take a stimulus—say, a word printed in a certain color—and ask subjects to name the color of the ink as fast and as accurately as they can. Now let the printed word be something tricky, punning on the task—the name of a color. J. Ridley Stroop apparently had some kind of fetish for punning (MacLeod, 1991b), and, of course, punning with words and images was a favorite sport in the visual arts around the time—practiced by, among many others, Marcel Duchamp and René Magritte (*La trahison des images*, "The treachery of images," or the famous painting of a pipe that said it was not a pipe, dates from 1928–1929). Punning was in the air and of the zeitgeist in the early 1930s.

So we design stimuli, words, printed in a color that either matches with the semantics of the word itself (the word RED printed in red) or does not (the word RED printed in green). Subjects invariably have a hard time focusing solely on the color of the ink. The semantics of the word inevitably kick in, distract, interfere, or interact with the relevant information processing. For congruent stimuli (sometimes also called "compatible" or "matching")—the word RED printed in red—subjects' responses tend to be faster and more accurate than for incongruent stimuli ("incompatible" or "mismatching")—the word RED printed in green.

The systems bias for the color of the ink turns out not to be strong enough and may leave us wondering whether the irrelevant semantics of the word facilitate the processing of congruent stimuli, inhibit the processing of incongruent stimuli, or both. To dissociate positive from negative effects, we can try to compare the performance in congruent and incongruent trials with a control condition, in which the semantics of the word would not prime a color (e.g., using words like CAT or BICYCLE) or in which there would be no semantics (e.g., a string of X's). Coming up with the appropriate control condition is not easy—a hardheaded cognitive psychologist will surely point out that cats and bicycles prime colors or that a string of X's in fact carries quite a lot of semantic connotations, not to mention the perceptual confound in the usage of a string of similar characters. Perhaps anagrams of the color words might be the best alternative, provided that the subject in question does not possess some fluke genius ability to solve anagrams within hundreds of milliseconds.

Psychologists have of course tried out several types of "neutral" stimuli. The gist of the entire data set, comprising hundreds of experiments over several decades of research (extensively reviewed and fairly assessed by MacLeod, 1991a), suggests that there are positive as well as negative effects. Performance in congruent trials tends to be a bit more accurate and faster than in neutral trials, suggesting that the word RED facilitates the color discrimination of red ink. In turn, incongruent trials tend to produce more errors and slower responses than neutral trials, pointing to a disturbing effect of the word RED on our ability to say that the color of green ink is green. In fact, the negative effects are generally slightly larger than the positive effects, but this could be skewed by a ceiling effect in congruent trials (it might physically be impossible to be even more accurate and faster than we already are in congruent trials, no matter how much extra push we get from irrelevant semantics).

Another notable phenomenon in the data with the Stroop test is the signature asymmetry of the effect, with irrelevant semantics of the word having an impact on color discrimination but not the other way around. When asked to name the word (or simply to read the letters that form the name of a color), subjects can generally do so without being bothered by the color of the ink in which the word is printed. Most likely, this is due to the strength of the association between stimulus feature and response (Washburn, 1994). Words are strongly associated with naming or reading, whereas ink colors are not. Put differently, word reading may be "overlearned" to the point where it has become automatic, requiring virtually no conscious effort—it would, in fact, be difficult to stop ourselves from reading a word that flashes before our eyes. Verbally labeling the color of ink, on the other hand, is not something we usually do. Our systems bias for bias will get us to read words but not name ink colors.

J. Ridley Stroop inexplicably, sadly, but probably also happily, abandoned the research in this area almost as soon as he had earned his doctoral degree over it, only to devote much of the rest of his life to very intense religious pursuit—his *opus magnum* being the self-published trilogy *God's Plan and Me*. Yet, the test he had devised continues to be a standard tool. The undiminished popularity of the Stroop test in psychology and the cognitive neurosciences must be thanks to how wonderfully simple yet rich in concept it really is. Over the years, many variants of the conflict paradigm have surfaced, with all kinds of stimuli, in any perceptual modality, but always revolving around the notion that stimuli can have multiple features with compatible or incompatible stimulus–response mapping (see Kornblum, Hasbroucq, & Osman, 1990, for a taxonomy). The paradigm is still one of the most suitable for researchers to assess how, and how well, a systems bias enhances relevant and suppresses irrelevant information (see Polk et al., 2008, for a case in point with BOLD responses).

The Stroop test initially focused on semantic interference from verbal stimuli—not exactly a linguistic kind of stimulus (devoid as it is of syntax) but one that might have briefly raised the notion that the interference effects depended on language. Yet, the conflict paradigm is, in fact, easily translated into tests with nonhuman species. My favorite example of cross-species translation would have to be a paper by Tang and Guo (2001) on "Choice Behavior of Drosophila Facing Contradictory Visual Cues." Tang and Guo developed a tiny flight simulator for the fruit flies and trained the little creatures to fly this way or that depending on either the color or the shape of a visual stimulus. The fruit flies were eager enough to perform the task, as they risked punishment for choosing the wrong fly path—"a beam of infrared light," "as an instantaneous source of heat" (p. 1546) of which the authors do not mention exactly how noxious it was (we can only assume that the subjects were not being fried alive, helplessly tethered as they were to the torque meter).

In one of the experiments, the fruit flies were trained to fly toward a green upright T and to avoid a blue inverted T at a particular level of color intensity. After training, the flies were presented with a test using new stimuli at different levels of color intensity: green inverted Ts and blue upright Ts that had either higher or lower color intensity than the stimuli used during training. Which stimuli would the fruit flies choose, given the acquired color and shape associations? If the fruit flies rely on color, they should choose the green inverted Ts; on the basis of shape, the preferred option should be the blue upright Ts (during this posttraining test, the fruit flies were not punished for either choice). The color intensity turned out to be a crucial determinant of the choice behavior. For color intensities higher than during training, the fruit flies made their choice on the basis of color. They relied on shape if the color intensities were lower than during training.

Most striking about these data was the rather sharp switch in the fruit flies' strategy of choice. Their systems bias went firmly in the direction of one visual dimension, either color or shape. At least that was the case for normal or "wild-type Berlin" flies. Tang and Guo (2001) repeated the experiment with mutant fruit flies, whose mushroom bodies were very small—mushroom bodies are neural structures of which the French nineteenth-century biologist Félix Dujardin (1850) had proposed that they endow the insect with a degree of free will. With not enough neural substrate of free will, the mutant flies proved to be more indecisive than their wild-type cousins, showing a less pronounced relationship between color intensity and the systems bias for either color or shape. To me, the entire paper by Tang and Guo (2001) reads almost like experimental poetry, highly intelligent, wonderfully innovative, and undeniably touched by a bit of the spirit of Gary Larson's cartoon series *The Far Side*. Nevertheless, the very serious conclusion must be that something like the Stroop test works for just about any creature with a nervous system and that research of this variety can be very informative indeed about an organism's adaptive control of information processing and decision making.

It is certainly also possible to devise Stroop-like tests for monkeys (Washburn, 1994). Using a manual go/no-go task with multidimensional visual stimuli, Sakagami, Hikosaka, several colleagues, and I explored the extent to which we could get interference effects from irrelevant features in task performance by macaques (Lauwereyns et al., 2000). We sampled four visual features—location, color, shape, and motion direction—using compound visual stimuli that looked like an object or aperture against a black background, inside which a random pattern of dots moved coherently in one direction. The contour of the aperture defined the shape (say, a circle or a plus). This aperture remained fixed in position, so that the motion feature pertained only to the dots moving inside. The dots were all in the same color (green or red for one stimulus set and purple or yellow for another).

The monkeys were trained to associate one feature within a visual dimension with a go response and the alternative feature with a no-go response (see also figure 6.2, panel a, with more abstract responses A or B). The discrimination tasks were blocked, so that the monkeys would have to perform a series of maybe sixty trials making their two-choice responses to the same visual dimension (say, color) throughout the block of trials. In the next block of trials, the task-relevant dimension might be motion direction, for instance. This arrangement reproduces the core feature of the Stroop test, with multiple stimulus–response mappings pertaining to the same visual object. There will be congruent trials, when the relevant stimulus feature primes the same response as an irrelevant feature. For instance, in figure 6.2a, in the color-discrimination task, the relevant color black primes the same response as the irrelevant downward motion direction. In incongruent trials, the stimulus–response mappings diverge. In figure 6.2a, in the color discrimination, this is the case for a black stimulus with upward motion direction.

Both with response time (in go trials, since no-go trials were not speeded) and proportion correct (in go as well as no-go trials) as the dependent measure,

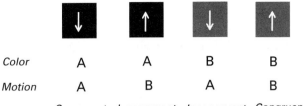

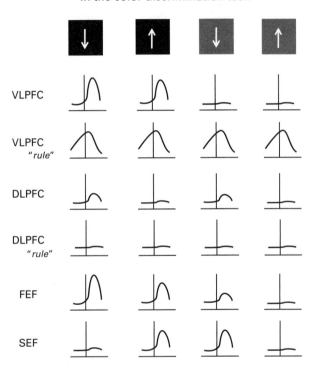

Figure 6.2
Stroop-like interference in the brain. (a) Two-choice task with compound visual stimuli, comprising a color (black or gray) and a motion direction (down or up). In the color-discrimination task (first row), one color is associated with response A and the other with response B, while motion direction is irrelevant. The roles are reversed in the motion-discrimination task (second row). This schedule produces two congruent (AA and BB) and two incongruent stimulus combinations (AB and BA). (b) Neural responses in different brain areas during the color-discrimination task for each of the four stimulus configurations. The vertical line in the middle of each graph represents the time of stimulus onset. Responses are shown for two classes of neurons in the ventrolateral prefrontal cortex (VLPFC) as well as the dorsolateral prefrontal cortex (DLPFC). Responses are also shown for neurons in the frontal eye field (FEF) and the supplementary eye field (SEF).

we observed clear interference effects from irrelevant features for all but one type of visual discrimination (Lauwereyns et al., 2000). When the monkeys were asked to discriminate the spatial position of the visual object, they did so very fast, making very few errors, with no influence to speak of from irrelevant color, motion direction, or shape. The data were in agreement with the proposal that visual selection by position is something "special," performed at a very early stage of processing, allowing the cognitive system to disregard any further information contained by the stimulus (Theeuwes, 1994; Van der Heijden, 1992)—a proposal that is also consistent with the notion that the speed of processing has a critical impact on whether irrelevant information produces interference with task performance. Processing position is so easy, and can go so fast, that everything else remains untouched in the background.

In color-, shape- or motion-discrimination tasks, on the other hand, the monkeys' performance was markedly better in congruent trials than in incongruent trials. As we did not employ a neutral condition without any acquired stimulus–response mapping, we had no means to check directly whether the interference was due to facilitation in congruent trials and/or inhibition in incongruent trials. With MacLeod (1992), we simply assumed that a bit of both would be at play. In our data set, we thought another type of analysis might be more interesting. Since the stimuli consisted of more than two stimulus features for some of the monkeys, we were able to test whether the effects from multiple irrelevant features were superadditive.

This type of analysis would tell us whether effects from different irrelevant features were independent of each other, one disturbance effect simply adding onto another, or whether there would be an interaction. The data suggested the latter, with a pronounced nonlinear effect, implying a level of disturbance with two incongruent features that was more than twice as much as that with a single incongruent feature (Lauwereyns et al., 2000). From the perspective of bias versus sensitivity, this superadditive effect suggested something like a multiplicative scaling effect, consistent with the notion that performance was primarily affected by a changed signal-to-noise ratio. The more noise, or the more irrelevant features, the smaller the signal-to-noise ratio. Unfortunately, we did not have sufficient data to apply the LATER model in our analyses. Yet, the nonlinear relation between the signal-to-noise ratio and response time does in and of itself already seem to confirm that the behavior was negatively governed by sensitivity mechanisms—not by an improvement of sensitivity as a function of reward-oriented, top–down, or strategic information processing but by a deterioration of the signal-to-noise ratio by the excess of irrelevant information in the given situation.

As a corollary to this interpretation, the interaction among multiple irrelevant features also suggests that the interference occurred at a level of information processing that integrates information from different visual dimensions—either a rather abstract level of decision making or a level of processing oriented already toward motor preparation. In any case, it seems fair to say that the superadditive effect suggested a weighing of decision options done in a currency that values sensory evidence from several sources. It certainly reflected an interactive competition among learned stimulus–response associations—associations that were entirely arbitrary to begin with and would have come to the monkey's mind only because they meant something specific in the behavioral context (in this case, a drop of orange juice for a correct manual response).

Having established the usefulness of the behavioral paradigm as a monkey version of the Stroop test, we were able to move on to single-unit recording in a search for neural correlates (Lauwereyns et al., 2001; Sakagami et al., 2001). With color and motion direction, we focused on just two visual dimensions, maximally segregated in terms of the underlying neural circuits—color being primarily associated with the ventral stream of visual information processing, the so-called "what system," and motion direction as a prototype stimulus for the dorsal stream, the "where system" (referring to a well-known and much debated distinction first made by Ungerleider & Mishkin, 1982, and probably suffering from excessive simplification, being a form of overeager abstraction based on fact rather than fiction; see Felleman & Van Essen, 1991, for a more intricate wiring of the monkey brain).

With associations between color stimuli and manual responses, as well as between motion stimuli and manual responses, our paradigm ensured that the task-relevant information processing relied entirely on context-dependent routes from sensation to action. It is a more adequate Stroop analogue than, say, the antisaccade paradigm (Munoz & Everling, 2004), which also generates conflict as a function of stimulus–response compatibility but does so between a "reflexive" or "hardwired" response (making a spatial response to the spatial position primed by a sensory stimulus) and a more "voluntary" one (making a spatial response to a different position in space). Instead, in our monkey Stroop test, the behavioral context dictated which visual dimension was associated with reward in a particular block of trials, allowing the systems to develop a reward-oriented bias, prioritizing one visual dimension (and its set of stimulus–response mappings) over another.

Figure 6.2b offers a stylized overview of the dominant findings in neural activity during decision making under conflict. The figure integrates the data of Sakagami et al. (2001) and Lauwereyns et al. (2001), especially with respect to the dorsolateral prefrontal cortex, the ventrolateral prefrontal cortex, and the frontal eye field. The figure also translates and incorporates data from several cognate studies that provided additional clues—particularly Stuphorn, Taylor, and Schall (2000) and Nakamura, Roesch, and Olson (2005) with respect to the supplementary eye field and Wallis, Anderson, and Miller (2001) with respect to prefrontal neurons that appear to encode abstract rules.

Shown are the neural firing rates in the color-discrimination task, averaged to the ideal of an infinite data set, in response to the four different types of stimuli reproduced at the top of panel b. The left- and rightmost stimuli belong to congruent trials, as the relevant color information primes the same response as the irrelevant motion information. The two middle stimuli are for incongruent trials. The panel presents six types of neuron, each row reflecting one type as it responds to each of the four stimuli. The thin vertical line in the middle of each of these four minigraphs indicates the time of stimulus onset. To the left of the middle line, we have half a second of neural trace before stimulus onset and, to the right, half a second after stimulus onset.

The first type of neuron (VLPFC, top row of figure 6.2b) is the most frequent in the ventrolateral prefrontal cortex and discriminates stimuli on the basis of their color (Sakagami et al., 2001). Here, the black color produces strong activation soon after stimulus onset (the left two minigraphs), whereas the gray stimulus does little to excite the neuron. Looking with a looking glass at the responses, we might detect a small and, in fact, statistically reliable difference between the black congruent trial (leftmost minigraph) and the black incongruent trial (second minigraph from the left), with a slight reduction of activity in the incongruent trial (Lauwereyns et al., 2001). The second type of neuron, also found in the ventrolateral prefrontal cortex (VLPFC "rule," second row), works mostly before stimulus onset, raising its baseline of activity toward the crucial onset of a task-relevant stimulus (Lauwereyns et al., 2001; Sakagami & Niki, 1994; Wallis, Anderson, & Miller, 2001). Both of these ventrolateral types of neuron, then, are in the business of signaling something relating to color. The first type carries discriminative information; the second presumably reflects a systems bias for color when color is deemed task relevant and a predictor of reward.

The third and fourth types of neuron are analogous to the first and second, but we find them in the dorsolateral prefrontal cortex (DLPFC), and they are mostly concerned with conveying information about motion direction (Lauwereyns et al., 2001; Sakagami & Tsutsui, 1999). The dorsolateral prefrontal "rule" type of neuron remains virtually silent in the color-discrimination task. Yet, the dorsolateral prefrontal discriminative neuron appears to leak some information about motion direction, with a small bump of activity for, in this example, downward motion directions (the leftmost minigraph and the third minigraph from left), even though motion direction is task irrelevant in the current situation (Lauwereyns et al., 2001). This leakage of motion information in the color-discrimination task could very well be the reason why other neurons, further down the path of decision making, get "Strooped."

The dorsolateral prefrontal neurons do for motion direction what the ventrolateral prefrontal neurons do for color. Thus, if we were to draw a stylized figure of the neural responses in the motion-discrimination task, we would find the reverse pattern, with proper motion discrimination and an anticipatory systems bias in the dorsolateral prefrontal neurons, silent ventrolateral prefrontal "rule" neurons, and, finally, ventrolateral prefrontal discriminative neurons leaking information about color when all the monkey should care about is motion direction.

The fifth type of neuron, found in the arcuate sulcus area, including frontal eye field (FEF, fifth row of figure 6.2b), is markedly different from the four earlier types. It discriminates among stimuli but is able to do this on the basis of both color and motion direction—Sakagami and Tsutsui (1999) dubbed this type of neuron the "CM cell." Its ability to integrate color and motion information suggests that it operates at a stage of information processing closer to the decision output than its more "single-minded" counterparts in the dorsolateral and ventrolateral prefrontal cortex. Perhaps most striking about the CM cell is the Stroop-like interference from irrelevant information evident in its firing rates (Lauwereyns et al., 2001).

In the example in figure 6.2b, we see a type of CM cell (FEF, fifth row) that increases its activity for black but not for gray stimuli. This distinction of black versus gray is very pronounced in congruent trials (leftmost versus rightmost minigraph) but not in incongruent trials (the two middle minigraphs). Put differently, the neuron's sensitivity is degraded under conflict, precisely as we should expect to happen somewhere in the brain, given the behavioral data. This pattern of activity would emerge naturally if the CM cell derives its information from dorsolateral and ventrolateral sources simultaneously. The compromised sensitivity in incongruent trials, then, would be due to the leakage of "noise" by neurons that fail to be entirely silent when the information that they code is presently irrelevant. Conversely, such irrelevant information would actually help exaggerate the CM cell's sensitivity in congruent trials.

The final type of neuron included in figure 6.2b (SEF, bottom row) is usually found in the supplementary eye field (Stuphorn, Taylor, & Schall, 2000) but sometimes also in anterior cingulate cortex (Nakamura, Roesch, & Olson, 2005). It signals the existence of conflict, firing more in incongruent trials than in congruent trials—as if it is able to detect a difference of opinion among dorsolateral and ventrolateral prefrontal neurons, or a lack of resolution in CM cells. This type of neural mechanism has also been documented extensively in fMRI research (see MacDonald et al., 2000, for a very clear distinction between activities relating to task instruction versus the presence of conflict). The jury is still out on whether the code of the supplementary eye field (SEF) type of neuron contributes to the monitoring of performance, signaling errors after the fact, or rather to something more online, like predicting the likelihood of success (and reward) or calling for extra effort and a tighter systems bias (as suggested by Pochon et al., 2008). However, the general agreement among researchers seems to be that this conflict signal is a crucial component of the adaptive control of information processing in the near future (in the next trial) if not immediately (in the current trial).

Thus, we have at least six types of neurons and a complex pattern of activity to consider. Quickly glancing at the entire panel b of figure 6.2, it can seem a daunting task to try to piece all these different signals together into a coherent picture of the neural circuits underlying Stroop-like interference in decision making. Listening to each type of neuron in turn is, of course, a fascinating exercise, but to move from contemplation to comprehension, the piecing together is exactly what we need to do. We must try to imagine the different signals as output from various elements at different positions within a neural network. Luckily, once we actually start drawing such a network, the hypothesized connections and information flow come with wonderfully precise properties and constraints that immediately prove to be compatible or not with the actual data. Using the data as our template, we may well home in on something sensible before we know it.

A Toy Model of Competition in the Brain

Inspired by connectionist models of decision making under conflict such as those by Cohen and Servan-Schreiber (1992), Zhang, Zhang, and Kornblum (1999), and Botvinick et al. (2001), I propose a little toy model of competition in the brain that provides a plausible layout of the six types of neuron identified in figure 6.2 (see Lauwereyns et al., 2001, for an earlier version). Figure 6.3 presents the core architecture of the toy model. Figure 6.4 depicts the activity pattern on a congruent trial in the color-discrimination task, whereas figure 6.5 shows what happens on an incongruent trial in the same task. Though I have not attempted an actual computer simulation of the data (I gladly refer to the three bona fide modeling papers above for efforts of that kind), I humbly suggest that all I have done, really, is to redraw extant models with labels that pertain to the monkey Stroop paradigm with color versus motion discrimination. I think none of the features of my toy model are invented by me or peculiar to figures 6.3, 6.4, and 6.5—in fact, I have tried to draw the simplest possible model that matches the data, employing the most uncontroversial properties of the three referenced papers. Thus, properly speaking, I believe this is not even my own little toy model, but, at least in spirit, that of Jonathan D. Cohen and like-minded computational neuroscientists.

The basic flow of information, from sensory input to decision output, works from left to right in figure 6.3, with two parallel streams of visual processing: the dorsal stream (the upper modules) in charge of motion direction, with

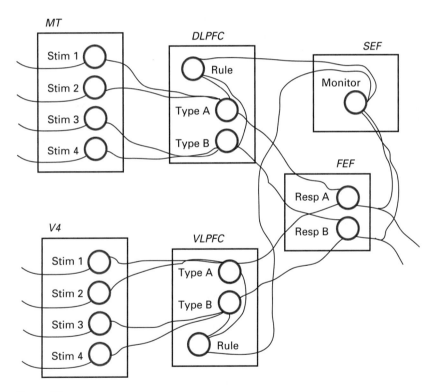

Figure 6.3
Schematic representation of the neural network directly involved in Stroop-like interference. Rectangular shapes indicate neural modules that consist of groups of neurons with different response properties (each dot represents one group). The wires indicate the dominant connections among neurons from different neural structures. Shown are the following brain areas: MT, middle temporal area; V4, fourth visual area; DLPFC, dorsolateral prefrontal cortex; VLPFC, ventrolateral prefrontal cortex; FEF, frontal eye field; SEF, supplementary eye field; Stim, Stimulus; Resp, Response.

area MT and the dorsolateral prefrontal cortex, and the ventral stream (the lower modules) in charge of color, with area V4 and the ventrolateral prefrontal cortex. The visual cortical areas MT and V4 each contain neurons that are tuned to specific features within the visual dimension—so MT neurons are tuned to particular motion directions, and V4 neurons to particular colors. In this toy model, we consider only four possible motion directions and four possible colors (Stim 1–4). These visual neurons project their information in a special partnership with a particular prefrontal cortical area. Importantly, at the level of prefrontal cortex, the sensory information is already "interpreted" or "classified" as a function of behavioral context. Thus, dorsolateral prefrontal cortical neurons divide into two discriminative nodes, one representing Response (or Type) A and the other Response B. These nodes are activated by the associated motion directions—via connections acquired by fixed stimulus–response mapping. In the example, Stim 1 and Stim 2 project to Type A, whereas Stim 3 and Stim 4 project to Type B. An analogous architecture exists between area V4 and ventrolateral prefrontal cortex in the ventral stream.

Still in dorsolateral and ventrolateral prefrontal cortex, we also find the "rule" neurons, which are connected with the discriminative nodes in their respective modules, as well as with the monitoring node in the SEF. In the next level up in the hierarchical organization of decision making (or the next level toward motor output), we have two response planning nodes (Resp A and B) in the arcuate region of prefrontal cortex, including the frontal eye field (FEF). Resp A and B are activated by the associated discriminative nodes in the previous layer—so both the dorsolateral and the ventrolateral prefrontal Type A project to Resp A (and Type B to Resp B). Finally, the Resp A and B units project to motor areas to initiate the behavioral response, but they each also send collaterals to the monitoring node in SEF.

For simplicity, I have drawn only excitatory feed-forward projections (except the connections between SEF and the "rule" neurons, which go in the opposite direction, from right to left, from a "higher" to a "lower" module). I would like to think of the core architecture in figure 6.3 as one that represents the current state of affairs, after extensive training, in the underlying anatomy that contributes to task performance in the monkey Stroop test. Connections have been acquired, or have been given a certain weight, through positive reinforcement (say, fruit juice as a reward for correct responses). Now this basic architecture puts boundary conditions on information processing and decision making in the monkey Stroop test.

Figures 6.4 and 6.5 then show the ad hoc pattern of activity within the toy model in a given situation, in a particular trial. In these figures the thickness

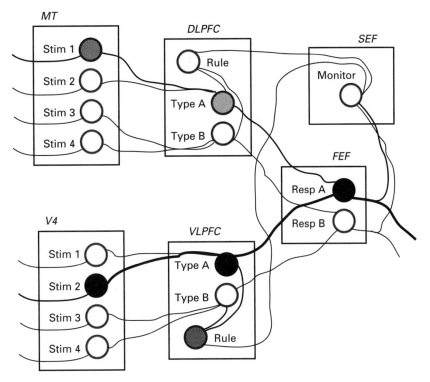

Figure 6.4
Pattern of activation in a congruent trial during the color-discrimination task. The pattern is shown for the neural network presented figure 6.3. The filling in of the dots and the thickness of the wires indicate the degree of activation: the darker the dots, the stronger the activation; the thicker the wires, the stronger the activation. MT, middle temporal area; V4, fourth visual area; DLPFC, dorsolateral prefrontal cortex; VLPFC, ventrolateral prefrontal cortex; FEF, frontal eye field; SEF, supplementary eye field; Stim, Stimulus; Resp, Response.

of the lines and the shading of the nodes indicate the level of activity—the thicker the lines, and the darker the nodes, the stronger the activity. Both figures show a snapshot of activity in the color-discrimination task. In figure 6.4, the monkey is presented with a compound visual target, comprising motion direction 1 and color 2. The task-relevant color strongly activates the corresponding V4 unit as well as the ventrolateral prefrontal Type A unit, but the irrelevant motion direction also gets MT unit Stim 1 activated to some extent, along with the associated dorsolateral prefrontal unit. At the next level up, the FEF unit Resp A is doubly activated, both by relevant information streaming in via ventrolateral prefrontal cortex and by irrelevant

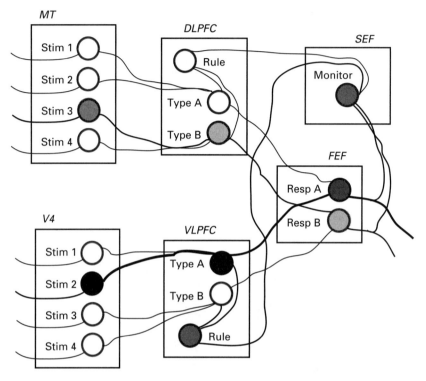

Figure 6.5
Pattern of activation in an incongruent trial during the color-discrimination task. The neural network and the presentation format are the same as in figure 6.4. MT, middle temporal area; V4, fourth visual area; DLPFC, dorsolateral prefrontal cortex; VLPFC, ventrolateral prefrontal cortex; FEF, frontal eye field; SEF, supplementary eye field; Stim, Stimulus; Resp, Response.

information originating from dorsolateral prefrontal cortex. There is no doubt in the monkey's mind, SEF stays asleep—the response should be A. The response time will be short and the likelihood of error very small.

Figure 6.5 tells a different story. Now the monkey is presented with a trickier visual target, comprising motion direction 3 and color 2. On the ventral side, things look the same as in figure 6.4. Again the task-relevant color strongly activates the corresponding V4 unit and its partner in ventrolateral prefrontal cortex (Type A). However, this time the irrelevant motion direction gets MT unit Stim 3 somewhat activated, which in turn wakes up the dorsolateral prefrontal Type B unit. At the next level up, the FEF units contradict each other. Resp A, activated by relevant information from ventrolateral prefrontal cortex, vies for a different response plan than Resp B, activated by irrelevant information leaked through by dorsolateral prefrontal cortex. What is the monkey to do? The concurrent and conflicting collaterals activate the monitoring node in SEF, which may try to reset or strengthen the systems bias for the relevant visual dimension (in this case, color). This might prove useful especially in the next trial, but for now the response time is likely to be stretched out and the likelihood of error may be high enough to be considered "considerable."

This toy model of competition in the brain incorporates both positive and negative interference effects from competition. The disturbance effect of conflicting information, I trust, is obvious enough. However, facilitation can occur as well, when relevant and irrelevant information converge on the same action plan, creating a larger push than what might derive from relevant information alone (say, with static color stimuli, or color stimuli with a motion direction that is not mapped onto any response, in the color-discrimination task). Arguably, this kind of facilitation corresponds to the benefits from competition as discussed with the example of Michael Phelps and his string of world records.

The model also accommodates the dimensional overlap theory by Kornblum, Hasbroucq, and Osman (1990), which emphasizes similarity as a crucial determinant of interference in stimulus–stimulus and stimulus–response compatibility. We could, in fact, go one step further, tentatively proposing the "neural overlap theory" by which similarity of information would imply the usage of the same neural circuits. A promising prediction then, going beyond the original dimensional overlap theory, would be that interference between relevant and irrelevant information could occur even when the different stimulus or response features do not share any obvious characteristics in a computational sense but happen to elicit neural processes in partially overlapping circuits (Fias, Lauwereyns, & Lammertyn, 2001; Pinel et al., 2004). This might be the case, for instance, with numerical information and the spatial orientation of a visual target (Lammertyn, Fias, & Lauwereyns, 2002), since both types of stimuli engage the dorsal stream of information processing. In terms of the toy model of figures 6.3 to 6.5, the information of spatial orientation and number would converge onto the dorsolateral prefrontal cortex, creating competition there that would not be the case between, say, color and number, since color information projects onto the ventrolateral prefrontal cortex. In this way, our toy model moves from being merely descriptive to making a few predictions of its own—a sine qua non for science.

Uniting with Language-Demons

Can the scientific investigation of competition in the brain take us another step further? If we are to believe Daniel C. Dennett: yes. In 1991, he notoriously claimed to have succeeded where countless philosophers and scientists had failed before him over thousands of years of thinking and experimenting—now the self-congratulatory title of his otherwise brilliant book promised that the mystery of body and soul was a thing of the past, the brain–mind riddle solved, *Consciousness Explained.* The secret of Dennett's success, or megalomania, or hundreds of pages of success and two title words of megalomania, lay in the proposal of a pandemonium model, in which consciousness and the voluntary control of decision making are taken to arise from the infighting among a crowd of little demons, neural processes, thoughts (Dennett, 1991, p. 275):

> According to our sketch, there is competition among many concurrent contentful events in the brain, and a select subset of such events "win." That is, they manage to spawn continuing effects of various sorts. Some, uniting with language-demons, contribute to subsequent sayings, both sayings-aloud to others and silent (and out-loud) sayings to oneself. Some lend their content to other forms of subsequent self-stimulation, such as diagramming-to-oneself. The rest die out almost immediately, leaving only faint traces—circumstantial evidence—that they ever occurred at all.

The spawning of continuing effects here of course refers to the possibility of a mental life, quite separate from the simple selection of an action, go or no-go, left or right. The competition produces a winner that gets to play a part in the virtual reality of imagined futures, contemplated possibilities, anticipated outcomes to a set of alternative, as yet unperformed actions. In our toy model of the brain, we would have to add some kind of loop, something recursive, Hofstadter's (1999) Eternal Golden Braid, a mechanism that would generate a more elaborate decision-making process, double-checking or reevaluating sensory evidence but also performing more complex computations and abstractions up to the development of scenarios with a labyrinthine structure of clauses and conditions thought up by our relentless language-demons ("Taking *this* into account, if *that* happens, we might be better off doing *the other thing*, unless…").

The topic of brain, language, and consciousness asks for a different book—a few good ones have been written already (I particularly like *The Symbolic Species* by Terrence W. Deacon, 1997; along with Steven Pinker's, 2000, must-read *The Language Instinct*; and Ray Jackendoff's, 2007, *Language, Consciousness, Culture*), but the writing for most of the good ones, I hope, has yet to begin. Here, I will zoom in on a much simpler question. I would like to focus on the idea that consciousness, the spawning of continuing effects in cognitive processing, or the extra loop in decision making, can be said to disengage from immediate behavioral response. Instead of flexing or extending muscles right away, the conscious decision maker adds an extra round of thought to the process. For this literal form of transcendence to occur, for an unconscious tendency, an implicit bias, or, as it is sometimes called, a "prepotent" response, to be countermanded, interrupted, reexamined, or otherwise resolved, we need a certain something that either stops what is about to unfold, encourages a different evolution of events, or both. How does it work? How do we escape, how do we free ourselves from bias?

7 Utopia—A World without Bias

"Here we are"—You can't
hear us without having to be
us knowing everything we

know—you know you can't

Verbal echoes so many ghost
poets I think of you as wild
and fugitive—"Stop awhile"

These spare and impossibly demanding words are borrowed from Susan Howe's *Souls of the Labadie Tract* (2007, p. 58), a lyrical investigation, "[a]rmed with call numbers" for Yale's Sterling Library, where the poet found her way "among scriptural exegeses, ethical homiletics, antiquarian researches, tropes and allegories, totemic animal parents, prophets, and poets" (p. 15) following the traces of the Labadists, whom the back cover of the book informs us were "a Utopian Quietest sect that moved from the Netherlands to Cecil County, Maryland, in 1684" only to dissolve in 1722. To think, we must stop awhile, the poem advises. As we heed the verbal echoes, we must realize that we cannot *hear* them—sure we may hear *something*, but unable as we are to become the other, you, the one who spoke the words, we are left with a few beautiful noises, wild and fugitive, and the task of figuring out where they came from and where they will go, even as they claim they are *here*.

The reference to Utopia further accentuates the hopelessness of the task, the conjunction of *u-* and *topos* creating a Greek neologism for "No-Place." The word was coined by Thomas More (2003) in his secondhand report of a distant island where the inhabitants tolerate difference in religion and know of no private ownership of land. Hythlodaeus ("Dispenser of Nonsense") was the man said to have visited the strange place. More described how he met the explorer in 1515 in Antwerp (my hometown in Belgium). The distant island was to be situated far beyond Brazil, presumably beyond the then unnamed Cape Horn. Perhaps the great fantast had sailed via the then unnamed Mar de

Hoces, also known as Drake's Passage, all the way to the then unknown islands of New Zealand, or Aotearoa, the Land of the Long White Cloud (where I live). With a little stretch of the imagination the North Island fits the descriptions. To complete the circle, More first published his *Utopia* in 1516 under the editorship of Erasmus in Leuven, or Louvain (the city of the university where I obtained my doctoral degree).

No-Place, accepting different gods and encouraging the thought that everything belongs equally to everyone, would be a world without bias alright, but Thomas More, combative Catholic, ardent persecutor of Protestants, and wealthy landowner, surely did not think of it as perfect or ideal, and scholars are still arguing about the extent to which *Utopia* should be seen as a satire, or rather as food for thought, a philosophical experiment, an exercise in conceptual differentiation, stopping awhile to reflect on the sounds of other voices. I, for one, am inclined (biased) to see the best use for contemporary readers in *Utopia* as a roundabout effort to make us aware of the impossibility of escaping bias, subjectivity, and personal investment. Reading the book, I cannot prevent myself from expanding on its peculiarities and anecdotes with private associations—this overzealousness in the semantic department, I plead guilty, would be a disease I readily suffer from. The symptoms surface inevitably when I read novels, poems, monographs, scientific articles, anything that engages my cognitive system strongly enough to continue reading. In the case of *Utopia*, the effect resonates more than elsewhere because the private associations echo what I see as the main theme of the work. They put further emphasis on the fact that I cannot be anybody but me with all my identifying features, all my wants and needs, my habits and expectations, in whatever I do or think.

Take Two

If I cannot escape me, myself, and I, then what might I imagine a world without bias to be like? The question is absurd. Without fears and desires, without preferences for good over bad, there can be no "I," no "you," no "we." Everything would be deprived of meaning. Nothing would matter, and each action or event might as well go this way or that, completely random, chaotic—the way things would have been in the primordial soup of simple organic compounds from before life on earth began. Utopia would be any place in the universe (or multiverse) where no life forms exist—by this definition, the vast majority of extraterrestrial space qualifies. In fact, we have yet to find anything but Utopia beyond our planet, though Fox Mulder, the fictional investigator of strange cases in the TV series *The X Files*, believes alien biases to exist somewhere out there in space.

For us, living creatures on planet Earth, the utopian quality of Utopia is probably not something to be sad about. From the abundance of data presented in this monograph, the one thing that emerges loud and clear is that bias in neural processing is so basic as not to be got rid of or ignored. In just about any cognitive activity, we find forms of additive scaling in anticipatory processing, accompanied by, or competing with, multiplicative scaling in synergistic processing. Signal detection theory and the LATER model become nonsensical as tools to analyze decision making if we try to remove the notion of bias, a movable criterion, or a given distance from a starting point to a threshold. Perfect equanimity or complete neutrality can exist only insofar as things do not concern us, when we have no vested interest, when outcomes do not affect us. In chapter 2, "Wish Come True," and chapter 3, "Fear Materialized," much of the discussion was premised on the notion that bias and sensitivity imply neural mechanisms that were shaped as a function of environmental selection pressures, through adaptation, exaptation, or any other creative or formative process, promoting inclusive fitness and evolutionary success.

Nevertheless, having human individuals plodding away, choosing actions, and making decisions for personal gain, or out of idiosyncratic conviction, comes with a peculiarly "modern" risk, pitted against the very core of inclusive fitness and evolutionary success. The risk was phrased most eloquently by the great philosopher and logician Bertrand Russell, for BBC radio on December 30, 1954 (MacArthur, 1999, p. 264):

> In the great world of astronomy and in the little world of the atom, Man has unveiled secrets which might have been thought undiscoverable. In art and literature and religion, some men have shown a sublimity of feeling which makes the species worth preserving. Is all this to end in trivial horror because so few are able to think of Man rather than of this or that group of men? Is our race so destitute of wisdom, so incapable of impartial love, so blind even to the simplest dictates of self-preservation, that the last proof of its silly cleverness is to be the extermination of all life on our planet?

Bertrand Russell spoke of the hydrogen bomb with a genuine fear, prominent in the minds of many in the 1950s and 1960s, a fear that some five decades later has curiously waned, habituated as we have become to cynical euphemisms and the political necessity of "missile defense systems." The call for impartial love and universal thought somehow has a sentimental ring to it, more utopian in the present day than it ever was. Yet, no one with a human heart and a healthy brain can fail to understand the timeless truth of the point that the philosopher wished to make. The well-being of the many requires that individual biases be held in check. There must be constraints—or put positively, degrees of freedom—that define the range within which biases can operate. At some point a "rogue bias" must meet a wall, a boundary, a limit beyond which it cannot tread without being subjected to some kind of counteractive or corrective mechanism.

How do we define such borders? How can we control individual biases so that they are used only advantageously, in agreement with the well-being of the many, and remediated when they imply danger, distress, or any other negative effect for others? If we cannot do without bias, if individual biases are intrinsic to the species that we are, then the crucial question must be how to negotiate among biases—how to manage differences in opinion, opposite tendencies, conflicting desires and fears. No doubt the first step in this process must be the identification of a problem or a conflict. We need a signal that tells us when it is time to stop awhile, to take two (seconds, minutes), to consider a corrective action, a second approach, a take two, a solution to the problem, a way in which we can resolve the conflict. Figure 7.1 illustrates an arrest of this kind in full progress.

Figure 7.1
Decision making on the road. The occasional hiker, stopping awhile, finds herself struggling with the notion that the path taken might not be the right one. Fatigue, the beauty of the local scenery, and the uncertainty of what lies ahead have conspired to request a take two. The subject, six years old in the meantime, is the same as the clamoring one in figure 3.3 and the contented one in figure 4.1. She is also the artist who produced the work shown in figure 5.1.

In the previous chapter, "Less Is More," I have already described neural signals that correspond to the process of problem identification. In Stroop-like stimulus–response compatibility paradigms, researchers have reported neurons in the supplementary eye field (Stuphorn, Taylor, & Schall, 2000) and the anterior cingulate cortex (Nakamura, Roesch, & Olson, 2005) that selectively increase their activity in incongruent trials—when there is a conflict between task-relevant and task-irrelevant information. The conflict signal can arise even before the subject receives feedback about the performance. Even if we would still like to consider this to be a mechanism of "performance monitoring," it is likely based on something more predictive than incorporating evidence that a given response was erroneous. The issue is important for the current discussion on how to manage potentially harmful biases—only *predictive* conflict signals create a window of opportunity to influence ongoing decision making.

Computationally speaking, a predictive conflict signal could easily emerge as a type of coincidence detection among competing "action plans," represented by different groups of neurons in a module for response programming, such as the frontal eye field in the case of eye movements. Coincidence detection is essentially what I implemented with the SEF type of neuron in the toy model of competition in the brain. Yet, there is good reason to think that this kind of signal belongs to a more general class predicting the likelihood that a particular response will lead to an error. Using a stop-signal task, Brown and Braver (2005) obtained evidence in BOLD responses that the anterior cingulate cortex predicts the likelihood of error even in trials without response conflict. This likelihood prediction was also independent of whether the subject actually made an error, exactly as we might expect from a predictive signal that allows us to prevent an impending disaster from taking place.

Kepecs and colleagues (2008) obtained evidence of another neural mechanism that appears to predict the likelihood of error in a task with only one relevant stimulus dimension. They recorded the activity of single neurons in the orbitofrontal cortex while rats performed a two-choice odor-discrimination task. The rats were presented with various odor mixtures and had to make a spatial response indicating which odor was dominant. The mixtures were either easily distinguishable as one of the alternative categories or more of a subtle blend, closer to the category boundary. Put differently, the stimuli implied different signal-to-noise ratios, with either clear or weak signals relative to the noise (in this case, the proportion of the alternative odor in the mixture). Kepecs et al. showed that the neurons tended to increase their firing rates in error trials, well before the rats received any feedback about their performance. This effect was particularly clear in error trials in which the odor mixture had a high signal-to-noise ratio. In this case, the rats would have been able to conclude on the basis of sensory evidence that they might very well have made a mistake, as if they realized, on second thought, that they had gone to the wrong choice port. The neural activity then seemed to correspond with a complete lack of confidence in the decision made. With the more ambiguous odor mixtures, the rats could hardly predict at all whether they had made a correct choice and would simply remain in the dark about their performance until they received the actual feedback.

In a behavioral experiment to follow up on the neural recordings, Kepecs and colleagues (2008) included a lengthy waiting period between the moment the rats categorized the odor and the time of reward (in case of a correct response). During this waiting period, the rats were given an opportunity to abort the current trial and reinitiate the task. The likelihood of reinitiation turned out to be particularly high in error trials—apparently the rats had a hunch, intuited, or otherwise realized when they had made a mistake just a moment ago, did not wait for the actual feedback, and decided to proceed with the next trial. The neural correlate of what the researchers called "decision confidence" translated into behavioral choice. When strongly in doubt, the rats decided it was wise to cut their losses, avoid the lengthy wait, and try better next time. Kepecs et al. argued that the combined neural and behavioral data pointed to a form of metacognition in rats, a fairly sophisticated type of anticipatory performance monitoring that provided a means for strategic control of behavior.

Neural signals of this kind provide a glimpse of how the brain supports the assessment of our own decisions and action plans. More generally speaking, it appears that highly evolved animal species such as humans, monkeys, and rats do more than simply gather sensory evidence and translate the perceptual representations to sequences of activity in skeletal muscles. We generate predictions on the likelihood of error and closely follow whether we are on the right track when we are about to commit to a particular decision. Signals of conflict or doubt improve our flexibility, our adaptability. They leave room for volition to intervene and change the course of mental and physical events. In humans, and likely to some extent in other species as well, this volitional moment triggers an extra round of cognitive processing that can include more abstract and complex representations of things and events, their defining properties, their economic values, and so forth. At some point consciousness and language may kick in, but the basic constituents of flexible behavioral control on the basis of metacognitive representations can already be observed in rats. It will be fair to assume that the architecture of the underlying neural circuitry in humans shows essential commonalities with that in monkeys, rats, and other mammalian species.

The anticipatory performance monitoring, and the estimates of uncertainty it implies, can also be regarded as an essential component of decision making when multiple brain systems promote different behavioral choices (Daw, Niv, & Dayan, 2005). This kind of choice problem is somewhat analogous to what happens in the brain during the Stroop test and can be seen, for instance, under "motivational conflict" (Watanabe, Lauwereyns, & Hikosaka, 2003a), when short-term versus long-term gains are in opposition, or when well-learned or habitual behavioral responses, coded in the caudate nucleus, go in a different direction from what ad hoc reward-oriented information processing in the dorsolateral prefrontal cortex suggests to be the better option (Kobayashi et al., 2007). If the first plan, the simplest or default plan, thinking in the short term or going by habit, bodes ill in given circumstances, the most swiftly effective solution to the danger of an erroneous choice will be to bypass it by activating a plan B, counteracting one bias by calling to arms another. As David Hume (2002, p. 266) formulated it: "Nothing can oppose or retard the impulse of passion, but a contrary impulse; and if this contrary impulse ever arises from reason, that latter faculty must have an original influence on the will." For Hume, the ability to activate plan B on the basis of projected error in plan A carries important philosophical implications. The influence of reason on the will may not exactly correspond to the age-old debate on whether there is such a thing as "free will," at least not in the sense that Schopenhauer (1999) gave to the concept in his landmark essay, written for a competition organized by the Norwegian Royal Society of Sciences in 1839. For the will to be truly free from anything worldly, unaffected by the prospect of pleasure or pain for self or others, it would probably have to operate in a utopian world without bias—back in the primordial soup. But Hume's comment sits well with contemporary notions of voluntary control, suggesting that to not go where a given passion leads you requires not merely error prediction but the activation of a counteractive mechanism, a bias against the bias.

One Bias against Another

Going back to the eye movement paradigm with an asymmetrical reward schedule (Lauwereyns et al., 2002b), discussed in chapter 2, "Wish Come True," we already have a framework for the study of neural mechanisms in opposition to bias. We can zoom in on what happens when the monkey is asked to perform an eye movement that runs counter to the reward-oriented bias. For instance, in one experimental condition visual stimuli presented on the left might be associated with a large reward, whereas stimuli in the opposite hemifield yield only a small reward. The trials with left versus right stimuli are randomly interleaved so that the monkey has no way of predicting where the next visual stimulus will be presented. The only thing for certain is the conditional rule: IF the stimulus appears on the left, THEN the monkey can earn a large reward by making a correct eye movement toward the position of the visual stimulus.

To ensure that the monkey will also perform the task in small-reward trials, we can apply a correction procedure. If the monkey fails to make the correct eye movement, the current trial will simply be repeated. This way, the monkey learns that there is no escape. Even in small-reward trials, the eye movement must go to the visual stimulus; otherwise, there will be no opportunity to proceed with the series of trials and no chance of earning a large reward in a subsequent trial. Strictly speaking, then, eye movements in small-reward trials are reinforced in two ways, with the current, less preferable reward, and with an increased likelihood (from zero to 50%) of a large reward in the next trial. In chapter 2, I focused on the mechanisms of reward-oriented bias and sensitivity to explain how the monkey's eye movements are facilitated in large-reward trials. However, this does not explain the entire pattern of behavioral data.

Comparing large- and small-reward trials with neutral trials from a condition with a symmetrical reward schedule (each spatial position being associated with an equal amount of reward, keeping the total amount of reward for the entire condition the same as for asymmetrical reward conditions), we found both facilitation and inhibition effects, with response times in neutral trials slower than those in large-reward trials but faster than those in small-reward trials (Watanabe, Lauwereyns, & Hikosaka, 2003a). Apparently, there was a behavioral cost associated with the task of working against the bias toward a large reward. There must be additional processes at play when the monkey detects a visual stimulus associated with a small reward. In this case, a motivational conflict arises, a double bind, in fact, as the monkey is forced to do voluntarily the thing that does not match with what—metaphorically speaking—the heart desires.

Figure 7.2 sketches a type of neural activity in monkey caudate nucleus that appears to reflect such a counteractive process (based on data from Watanabe, Lauwereyns, & Hikosaka, 2003b). The shaded area in panel a reproduces the anticipatory (large-)reward-oriented bias seen in other types of caudate nucleus neurons—this is the target activity to be opposed. Concentrating on caudate neurons that were mainly active after the onset of the visual stimulus and during the performance of the eye movement (close to half of the presumed medium-spiny GABAergic projection neurons we encountered during extensive probing), we noticed a sizeable minority of neurons whose firing rate

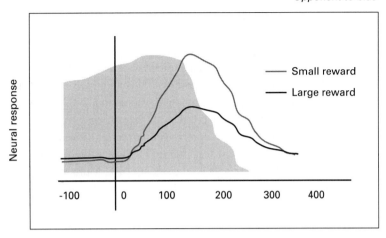

a

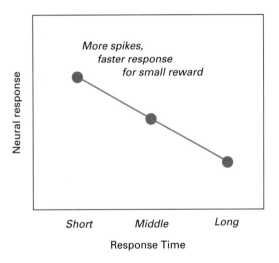

b

Figure 7.2
Neural correlates of a mechanism to counteract bias (reported in a different format in Watanabe, Lauwereyns, & Hikosaka, 2003b). (a) Neural activity in monkey caudate nucleus following a visual instruction (presented at time zero, indicated by the vertical line) to make an eye movement associated with a large reward (black line) or a small reward (gray line). The shaded area represents an anticipatory bias of the type introduced in figures 1.8 and 2.3. When the visual instruction conflicts with the anticipatory bias, in a small-reward trial, the neural firing rate is much stronger than when the instruction matches with the bias, in a large-reward trial. (b) Relation between neural firing and response time (short, middle, or long) in small-reward trials. The data suggest a facilitation effect of the caudate nucleus activity on eye movements for a small reward, with higher firing rates leading to shorter response times.

increased particularly in small-reward trials, that is, when the monkey was faced with the task of aborting any plan to make an eye movement in the direction associated with a large reward and instead had to find the energy to move in the opposite direction.

As such, the finding that a minority of neurons fired more for a large reward than for a small reward was not all that surprising and nothing really new (see Kobayashi et al., 2002; Shidara & Richmond, 2002). However, comparing the firing rates of these neurons with the response times, we obtained evidence that the neural activity contributed *positively* to the performance of an "underpaid" eye movement (see figure 7.2b)—the higher the firing rate, the more efficient the progress with the tedious task. More technically, if we divide the entire set of data from small-reward trials in three groups as a function of response time (short, middle, or long), we find that the firing rates are the highest when the monkey's action is swift, and they are the lowest when the monkey takes a long time to do the unappealing work that cannot be avoided.

The findings in monkey caudate nucleus, of course, immediately raise the question of how the neural correlate factors into the anatomical architecture that supports the counteractive process. An important clue for this comes from a study by Minamimoto, Hori, and Kimura (2005), examining neural activity in the centromedian nucleus of the thalamus while monkeys performed a hand movement task with an asymmetrical reward schedule (see Lauwereyns, 2006, for discussion). In contrast with the caudate nucleus, the thalamic structure harbored a majority of neurons that increased their activity in small-reward trials. These neurons exhibited a temporal profile that, as in the case of the "counteractive" caudate neurons, formed a perfect antidote to the anticipatory activity of "biased" caudate neurons, with a burst of action potentials following the presentation of an undesirable visual stimulus. The data pointed to a central role of the thalamic structure in the gating of neural activity in opposition to response bias.

Minamimoto and colleagues (2005) further showed that electrical stimulation of the thalamic structure actually disrupted the hand movement in large-reward trials—direct evidence that the centromedian nucleus works antagonistically to response bias. Both anatomically and functionally speaking, the centromedian nucleus appears to form a complex with the parafascicular nucleus (Minamimoto & Kimura, 2002; Royce, Bromley, & Gracco, 1991), a thalamic structure that presumably does for oculomotor control what the centromedian nucleus does for skeletal muscle control. For simplicity, given the lack of data on the neural properties of these structures during eye movement control, I will take the centromedian nucleus neurons to represent the thalamic complex, leaving open for discussion—or better yet, experimentation—whether it is really the centromedian nucleus or rather the parafascicular nucleus that should be inscribed in the neural circuit for eye movement control against response bias.

The "counteractive" neurons in caudate nucleus may very well cooperate with the thalamic complex to facilitate the opposition to response bias by removing the default inhibitory projection from substantia nigra pars reticulata onto the complex. However, this would not suffice to activate the complex. I think we must again look to the conflict signals and anticipatory performance monitoring in areas such as the supplementary eye field and the anterior cingulate cortex (Brown & Braver, 2005; Nakamura, Roesch, & Olson, 2005; Stuphorn, Taylor, & Schall, 2000) for a likely source that triggers the opponent mechanism in the centromedian–parafascicular complex. Figure 7.3 gives a schematic representation of my best understanding of the neural connections that support the monkey in making an eye movement that provides less than hoped for. Panel a serves as a reminder of the basic neural circuit for reward-oriented eye movement (introduced in chapter 2). Panel b introduces the supplementary eye field and the centromedian–parafascicular complex as crucial players in the opposition.

Several questions about bias against bias are only now surfacing or beginning to be sharp enough in focus to elicit empirical study. Does the bias against bias require active suppression of the original bias? Can we find neural signals that directly inhibit presently inappropriate action plans? Or does it suffice to push in another direction, to employ energy for the initiation of a contrary impulse? Does this counteractive mechanism play a role in effort-related decision making as well, particularly when a subject chooses a demanding action for a large reward over an easy action for a small reward (Walton et al., 2003)? Is this kind of mechanism more independent of, or only indirectly governed by, dopaminergic processes? I am thinking again of the curious observation by Liu and colleagues (2004), mentioned in chapter 3, when blocking dopamine D2 receptor binding actually enabled monkeys to work better than ever in trials without immediate reward. Schweighofer, Tanaka, and Doya (2007) suggested we should look particularly to serotonergic mechanisms to understand how we can bring ourselves to go for difficult, effortful courses of action that require a lot of patience and thinking in the long term.

Beyond the present moment, the application of a bias against bias probably also has a transformative influence on the neural circuit that underscored the original bias. This is not so much an effect of deliberate planning in the long term but rather a matter of neural plasticity for future purposes, complementary to the dopaminergic mechanisms of reward-oriented learning, to attenuate

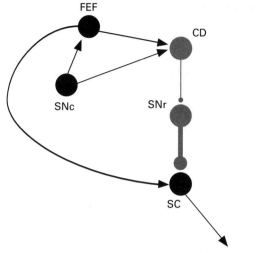

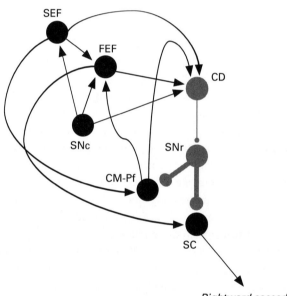

Figure 7.3
Schematic representation of a neural circuit to counteract bias in the control of eye movements. (a) For comparison, the basic neural circuit for eye movement control as presented in figure 2.4 is shown. Black arrows indicate excitatory connections; gray arrows indicate inhibitory connections. The thickness of the arrows represents the default strength of activity. FEF, frontal eye field; CD, caudate nucleus; SC, superior colliculus; SNr, substantia nigra pars reticulata; SNc, substantia nigra pars compacta. (b) Expanded neural circuit to show the relevant connections for processes in opposition to bias. The added structures are SEF, supplementary eye field; and CM-Pf, the thalamic complex including the centromedian and parafascicular nuclei. Performance monitoring in SEF exerts influence on CD, FEF, and PF to counteract the effect of anticipatory bias.

the first "impulse of passion" and the short-term thinking that it implies. If a particular reward-oriented bias must occasionally be kept in check, it pays to tone it down somewhat and prepare for more flexible control, with a healthy, that is, bounded, dose of wishful thinking that does not require excessive effort in recuperation and reorientation when events in the real world take a turn in the opposite direction.

Neural signals in the lateral habenula (Christoph, Leonzio, & Wilcox, 1986; Matsumoto & Hikosaka, 2007), which appear to derive from the globus pallidus internal segment (Hong & Hikosaka, 2008), show the exact characteristics we might expect from a process that has the power to retune the dopamine reward prediction error. The names of the neural structures seem to become more exotic as we try to map the circuit in greater detail—the lateral habenula ("little rein") is part of the epithalamus; the internal segment of the globus pallidus ("pale globe") belongs to the basal ganglia. We have more questions than answers about how the bias against bias really works and how an original bias gets modified, toned down, or, in the extreme case, abolished altogether and replaced by another. However, the good news is that the basic outline in functional and computational terms looks clear enough for concrete hypotheses and empirical investigation.

We Try to Overthrow Them

With the concreteness of hypotheses comes the possibility of another kind of bias, a researcher's bias, when we personally invest in our theories and conceptual schemes—a natural consequence of the central role given to authorship in the scientific community, with academic positions and status depending on publications and grants, exciting findings and persuasive new ideas. As we project our own future to be positive, and as the amygdala and rostral anterior cingulate cortex monitor the emotional salience of things in store for us (Sharot et al., 2007), it may be a tricky task to heed good epistemological advice when the results of our latest experiment start streaming in. It is easy for Popper (2002a, p. 279) to say the following: "Once put forward, none of our 'anticipations' are dogmatically upheld. Our method of research is not to defend them, in order to prove how right we were. On the contrary, we try to overthrow them." But high-profile cases such as that of Woo Suk Hwang, the disgraced cloning expert, prove just how tempting it can be to believe that we are on to "something important" when the empirical facts are not quite there yet to support the story. The world's first manufactured dog, Snuppy (Lee et al., 2005), still counts as a genuine achievement of the Hwang laboratory, but two groundbreaking papers on human embryonic stem cells (Hwang et al., 2004,

2005) suffered editorial retraction (Kennedy, 2006) when photos turned out to be bogus, data artificially inflated, and ethical protocols violated.

Somehow, we must marry the positive energy of optimism bias with the negative logic of Popper (2002a) for a sound and productive life in science—or even for a sound and productive life, full stop. The great epistemologist was, of course, absolutely right in hammering on the idea of falsifiability as the core principle in science. And perhaps falsifiability is just one form of a more general readiness to apply a bias against bias when the first inclination seems to be off the mark, in which case the vision of a fruitful tandem with positive energy and negative logic does indeed extend beyond matters of science.

As for science, the quote that I lifted from Popper's (2002a) *The Logic of Scientific Discovery* responds to Francis Bacon's (2000, p. 38) disparaging notes in *The New Organon*, Book One, aphorisms XXVI to XXX, on the "risky and hasty business" of our anticipations when we engage in philosophical thought about nature, human nature, and things as they are. These anticipations, Bacon wrote, "brush past the intellect and fill the imagination," prompting the conclusion that "[e]ven if all the minds of all the ages should come together and pool their labours and communicate their thoughts, there will be no great progress made in sciences by means of *anticipations*" (still on p. 38).

Popper (2002a) found himself completely at odds with this indictment. The anticipations that Bacon despised corresponded to what we would call hypotheses today, and though we might have a hard time in this or that case explaining by which kind of intuition, or inspired by what sort of anecdote, we come to harbor a particular image of how things work, it is equally hard to see how we could get started with any research unless we plan to look for *something*. The critical step is in the evaluation of predictions against outcomes. Optimism bias can be a useful ally in the generation of predictions and the planning of research and should be seen as one of a family of potentially positive "delusions," along with self-fulfilling prophecies and placebo effects, that can energize us, help us cope, or otherwise get us to think and behave in ways that actually do promote our well-being (Beauregard, 2007). The real issue is how to separate the optimism bias from the process of "interpretation" (as Bacon, 2000, called it). The data should have the absolute freedom to agree or disagree with the hypotheses.

For the data to have the power to arbitrate, we must create a context that is open and receptive to criticism. Karl Popper surely must have realized how *The Logic of Scientific Discovery* (2002a) depended on communication, glasnost, freedom of speech, and an environment that welcomes exchange, as he moved on from his 1935 masterpiece in epistemology to *The Open Society and Its Enemies* (2002b), his 1945 masterpiece in political philosophy, written in relative isolation in Christchurch, New Zealand, thousands of miles away from the catastrophic influences of totalitarian regimes in Europe. The key challenge in the establishment of an open society is easily identified as the submission of individual actions to review by others, with the well-being of the many in society as the final goal and the ultimate criterion for what may be deemed acceptable.

Thus, we must relinquish individual power and authority and somehow come to terms with the idea that to be part of an open society implies the assumption of responsibility and the readiness to be judged by our peers. That this challenge is, in fact, extremely difficult to meet is perhaps made most visible in miniature versions of real-life conflicts between individual benefits and those of the group. The best known example is the prisoner's dilemma, but other paradigms in game theory also implement a fundamental friction between the benefits for self versus those of the group, when the total payoff for all players is maximal in case everyone cooperates, but an individual stands to receive the biggest personal paycheck if he or she is the only one to defect, cheat, or choose the selfish option (see Camerer, 2003, for an overview of behavioral game theory).

Ideally, in Utopia, we would all be rational players, able to think dispassionately, neutrally, and objectively about which is the best solution for the benefit of our species, any species, the entire planet and beyond. Yet, even in the microworld of science, where rational thinking is exercised to the best of our abilities, we encounter highly intelligent players such as Woo Suk Hwang who take shortcuts for personal gain. On a much smaller scale, I bet many of us would have to admit moments in our scientific careers when we were not quite sure that what we were doing was "the right thing" but charged ahead anyway, conveniently forgetting about messy data from other experiments while writing up a research paper, or bypassing difficult ethical questions in the design of new experiments, satisfied with the idea that "Other people do it too," or "This is for the ethics committee to think about," even if that committee does not have the expertise to really understand all the ins and outs.

Luckily, the scientific community did develop methods of peer review that prevent most of the more suspicious turns in the interest of personal gain, and the publication culture further favors the quest for truths by stimulating the expression of comments, rebuttals, and difference in opinion. Faulty research and ill-conceived theories will get caught out at some point as the most persuasive ideas generally tend to be those that provide the soundest interpretation of a wealth of phenomena recorded by a legion of independent researchers. The individual biases then usually cancel each other out. The scientific community does not even need to resort to any excessive forms of punishment to achieve its open society—a damaged reputation is disastrous enough. The exposed cheat can seek a future outside science or apologize, clean the laboratory, and begin again, as in the case of Woo Suk Hwang, whose biggest offense might have been the failure to control against personal biases among his junior colleagues.

The Sentimental Duty of Punishment

If the scientific community achieved something like an open society with relatively benign control mechanisms, it is only because it can operate within a wider society that values the contribution of science and gives it a special status. Science can easily afford to lose unwanted people, but there is no such option for society as a whole. The dictates of a humane society offer us no choice but to continue living and working with each other, whatever our mistakes, misdemeanors, or crimes—the sanctions must of necessity occur within the context of society, even if these include the more radical types such as temporary removal in an impoverished environment (prison). In fact, for a society to function well and to generate productive levels of cooperation among its members, it will probably need to devise, or evolve, some kind of system that enables punishment of not only offenders but also people who fail to punish offenders (Boyd et al., 2003).

Negative responses to offenders appear to be a natural course of action for many of us, even in one-shot interactions with strangers we are likely never to meet again, and even if we incur some personal cost in the act of punishing. This phenomenon of "altruistic punishment," along with other forms of altruism, represents a powerful force in the organization of social relations and the balancing of personal biases. Nevertheless, there are certainly interindividual differences and contextual influences that modulate the levels of altruism seen in interaction, suggesting that it is not merely a matter of the human genome but also of cultural evolution and gene–culture coevolution (Fehr & Fischbacher, 2003). In the best of human societies we will acquire, at an age young enough to forget when it happened, the intuition to oppose an errant bias when we encounter one.

In an intriguing PET study (De Quervain et al., 2004), Ernst Fehr's research group added data on cerebral blood flow to the abstract concepts about human evolution. Subjects were invited to participate in a "trust game" (of the type introduced in chapter 2) in the role of first mover, that is, they got to decide how much money they would entrust their partner with, hoping to get a good share of the enlarged sum in return. Sometimes, however, the other player kept all the money—a clear act of selfishness, a norm violation, rated highly unfair by the subjects, who were then given the opportunity to punish the defector. The punishment could be symbolic or in hard cash, and when it was in hard cash, it either did or did not imply a personal cost for the subject. The analyses of cerebral blood flow focused on a one-minute deliberation period in trials in which the subject's trust had been betrayed by the other player. As the subject contemplated whether and how much to punish the unfair player, the blood flow to the caudate nucleus increased, especially when the subject considered docking hard cash from the other player's total, and even more so if this implied a personal cost for the subject. Given the well-established role of the caudate nucleus in reward-oriented control of action, De Quervain et al. hypothesized that the activation reflected "the anticipated satisfaction from punishing defectors" (p. 1254).

Of course, I like these data quite a bit, not only because they talk about one of my pet brain areas but also because they literally show one excessive reward-oriented bias (the unfair player's) being combated by another reward-oriented bias (the altruistically punishing subject's) in the service of justice and a healthy society. The concept of "altruistic punishment" should probably undergo a slight revision in the process, as the motivational component at work in the punisher does point to a very personal benefit in the form of *pleasure*. De Quervain et al. (2004) called it a "proximate mechanism," an emotional tool that effectively pushes people to do what rational thinking suggests would be for the good of all. The researchers compared it with proverbs such as "Revenge is sweet" (p. 1254), and I might add that the passionate punishing also rhymes with the peculiar joy some viewers experience when watching a movie like Quentin Tarantino's *Kill Bill* (particularly the 2003 volume 1), in which the heroine proceeds to eliminate her cruel adversaries in a rather very zealous campaign of righteousness.

The neural correlates of sweet revenge were examined further in an fMRI study by Singer and colleagues (2006), who added an empathic twist and more physical punishment to the experimental setup with the trust game. This time, subjects experienced the economic exchange with two other players in a preparatory stage before the actual collection of BOLD responses. As they made their first moves, the subjects saw their trust honored with a generous reward by one player and ruthlessly violated by the other. The subjects were also asked to rate the other players on several dimensions, including fairness, agreeableness, and attractiveness. Not surprisingly, players who disregarded the implicit social norm of answering trust by kindness were generally thought to be unfair, unpleasant, and perhaps even a bit ugly. Both of the other players were in fact confederates, that is, actors playing along with the experimenters to construct the right emotional decor for the critical second component of the experiment. In the second part of the experiment, both confederates as well as the subject underwent a behavioral procedure essentially amounting to torture—they each received electric shocks in turn.

Since the subject had to endure a few trials of painful stimulation as well, the researchers were able to examine the basic pattern of pain-related BOLD responses and then compare this pattern in trials when not the subject but one of the other players received an electric shock. This comparison provided a measure of empathic pain-related BOLD responses, or the extent to which the subjects were able to "feel the pain" of others. The data showed that both male and female subjects did empathize with "fair players," but the women felt somewhat less pain for "unfair players." Men were even more categorical—they felt virtually no pain for confederates who had betrayed them in the first stage of the experiment, and the activation of voxels in the nucleus accumbens suggested that there was indeed sweetness of revenge in this case.

In the studies by De Quervain et al. (2004) and Singer et al. (2006), the emotional processing around punishment of others was obtained using behavioral paradigms in which the subjects experienced direct, dyadic interaction with the other players, and so it seems reasonable to assume that the subjects felt some kind of personal involvement, even if there was nothing material to gain from whichever reckoning awaited anyone who deserved to be blacklisted and identified as "enemy." However, other human imaging studies show evidence of emotional processing also in moral decision-making tasks strictly limited to third parties, that is, other individuals that the subject does not know and has never interacted with—usually characters described in scenarios or people shown in a photograph accompanied by the briefest of descriptions.

Buckholtz and colleagues (2008) found that BOLD responses in the "rational" dorsolateral prefrontal cortex increased when subjects assessed, on the basis of different crime scenarios, whether alleged perpetrators should be considered responsible. The recognition of cause and effect looked like it could use some good old logical computation. In contrast, brain structures associated with emotional processing, such as the amygdala, the medial prefrontal cortex, and the posterior cingulate cortex, came into play when subjects pondered on the appropriate magnitude of punishment. Hsu, Anen, and Quartz (2008) reached a similar conclusion on the basis of fMRI data while subjects performed the somewhat cruel task of deciding how many meals to subtract from which poor and starving children. Apparently, we tend to behave as moral sentimentalists, relying on emotional processing to put a sense of fairness before strictly utilitarian reasoning on the basis of efficiency. To do the right thing, we need a good portion of well-aimed passion.

Arguably the most dramatic demonstration of the role of emotional processing in moral judgment was devised by Greene and colleagues (2001) in an fMRI study that presented subjects with difficult decisions about life and death for others, decisions that varied in the way in which they imply some kind of personal involvement. One example is the dilemma with a runaway trolley, unstoppable, headed toward five clueless people and their certain death. We can save all five good citizens if we turn a switch, and put the trolley onto another track, now headed toward only one poor victim. What should we do? Of course, we will feel bad about the one poor victim, but common sense leaves no doubt that one victim is better than five.

Let us consider a slightly different situation. Again we have a runaway trolley, and again it is headed toward a massacre with five dead bodies. We are standing next to a large male, unknown to us, on a footbridge above the track. This time we have no easy switch at our disposal, and the only way we can save the five people on the track is by sacrificing the big fellow, pushing him off the footbridge and on the track, where his massive body will block the runaway trolley. (We are too small and too skinny, so if we jump, the trolley will just keep going, making even more casualties.) What should we do now? Our first response may be uneasy laughter at the absurdity of the situation. But then, coming to our senses and probed seriously for a decision, most of us would say no. Somehow this situation looks much more like a crime, a heinous murder, compared to the simple business of just turning a switch. Yet, in both cases the choice is between saving one life and saving five.

We can try to analyze the two situations carefully and rationally to understand exactly how or why the moral judgments diverge, but the most salient difference arises intuitively and implicitly—one case evokes a fast and strong emotional response, whereas the other does not. Greene et al. (2001) obtained clear activation in brain areas associated with emotional processing for moral dilemmas that implied direct personal involvement. The activation in these areas was smaller for moral dilemmas with less direct personal involvement and was absent altogether for nonmoral dilemmas such as which is the best mode of transportation in given circumstances.

These demonstrations of a crucial role for emotional processing in moral decision making, eye-opening though they are, represent, of course, only a point of departure for a focused investigation of the underlying neural mechanisms and functional algorithms. The label *emotional processing* is about as vague as we can get in the description of internal representations of the positive or negative connotations carried by big concepts such as fairness, equity, justice, and their counterparts, unfairness, inequity, injustice. Now that we have proof of existence, we must explore the vague category of emotional processing with precise questions. Does the prospect of fairness act as a reward? Does it do so equally strongly when the prospect of fairness promises to be an attribute of "me" as opposed to "you" or "someone else"? Or is the emotional processing more salient in the form of loss aversion as when we dread losing our personal integrity? Does it involve truly internal reflection, or are we happy to lie and cheat as long as we do not get caught?

To tease apart the different forms and influences of emotional processing during moral decision making in various contexts, the concepts of bias and sensitivity could again prove to be extremely valuable research tools. Does the prospect of fairness effectively lower the threshold for one decision rather than another? Does it imply some kind of anticipatory processing before evidence streams in? Do we deliberate equally extensively on all the evidence, or do we give more weight to some types of information than to the rest? How well do we process the information; what is the signal-to-noise ratio?

We can try to devise behavioral measures of how much time subjects spend processing different types of information, and in which order. We can think of ways to record how decisions evolve over time, with prompts at different moments in the midst of deliberation to make an immediate forced choice. It should be easy to check how accurately the details of different cases are remembered and so on. These kinds of questions can be asked empirically in the current state of the art in cognitive neuroscience. To me, it seems the obvious next step in this line of research. By applying the concepts of bias and sensitivity, by looking for their neural signatures, we stand to gain a whole lot more knowledge about how we make moral decisions.

My Own Desire to Eat Meat

In the meantime, the data from cognitive neuroscience on moral decision making already carry important implications for philosophical and ethical perspectives, with the renewed emphasis on affect and subjectivity. The apparently inescapable "sentimental" dimension reinforces the idea that the computations of right and wrong, however complex and abstract, do in fact recruit neural mechanisms that have evolved as a tool for valuation of objects and events in a simpler metric of pain and pleasure—basic signals that allow humans, monkeys, rats, fruit flies, and probably just about any animal species to steer toward situations that are beneficial for survival and reproduction. These basic signals operate within the framework of one body, one individual, and one personal household and presumably correspond to the present state of affairs within that "self" as compared to some optimal condition, a balanced internal milieu, or homeostasis.

We may hope to transcend the individual perspective and apply the valuation of objects and events for the wider well-being of as many creatures as possible, but the basic human motivation to act, to do the right thing, is principally rooted in reasoning with emotions. Concepts such as neutrality and objectivity are relevant to moral decision making only insofar as they derive from an exchange among individual perspectives and personal affects—a symphony of biases. To be ready for such an exchange, to achieve it in the best way possible, we need to come to terms with the fact that we have a limited view, an inherently biased understanding of things.

There is no escape from the fact that whatever we think or feel takes place within a subjective perspective, whether we are indulging in private associations or honestly trying to approach an indisputable truth. In the words of Martin Heidegger, translated by Joan Stambaugh from part II of *Nietzsche* (2003, p. 29): "[W]hat constantly already lies present for representational thinking during representation which presents something to itself is the representer itself (*ego cogitans*) before which everything represented is brought, to which and *back* to which (*re-praesentare*) it becomes present." Perhaps being the great philosopher and the flawed human that he was, with a sharp intellect and the curious inability to recant obvious and disastrous mistakes, Heidegger understood more than most that the true nature of things can only appear, almost paradoxically, when individuals are able to assume the role of subject absolutely and completely. At least, I think that is more or less what he had in mind here (Heidegger, 2003, p. 60):

Object in the sense of ob-ject: Only when man becomes subject, that is, where the subject becomes the ego and the ego becomes the *ego cogito*, only where this *cogitare* is understood in its essence as "original synthetic unity of transcendental apperception," only where the culmination for "logic" is reached (in truth as the certainty of the "I think"), only there is the essence of the object revealed in its objectivity. Only there is it at the same time possible and inevitable to understand this objectivity itself as "the new true object," and to think it as unconditional.

To reach objectivity, to think "the new true object," requires a thorough understanding and a complete acceptance of the limits and constraints of the personal perspective. Such an understanding cannot be reached by a single person in isolation but might emerge as a product of mutual interaction among different individuals. It demands communication and negotiation among complementary, competing, and conflicting views—all in an effort to make implicit biases explicit, to chart the peculiarities of the subject, to understand exactly how the limits and constraints of the personal perspective operate.

Heidegger's notes on the role of the subject drive me back to the hidden "ego" in Descartes's Latin dictum, surely one of the most famous in the history of philosophy (already discussed in chapter 1). *Cogito ergo sum* was also used as something of a case study by Gilles Deleuze and Félix Guattari in *What Is Philosophy?*—on the creation of a concept, how it can be described as a an act of naming that works from an implicit base (Deleuze & Guattari, 1994, p. 61):

> Although Descartes's cogito is created as a concept, it has presuppositions. This is not in the way that one concept presupposes others (for example, "man" presupposes "animal" and "rational"); the presuppositions here are implicit, subjective, and preconceptual, forming an image of thought: everyone knows what thinking means. Everyone can think; everyone wants the truth.

Everyone knows what thinking means. Just like everyone knows what attention is. However, as soon as we verbalize the presuppositions, we realize that they are in fact ill-defined, that they should suffer explicit investigation. What is "thinking"? Who or what does the thinking? What Deleuze and Guattari see as the *philosophical* task of creating concepts boils down to identifying important implicit information that needs to be addressed in broad daylight for a full and proper understanding of phenomena. With Heidegger, I would like to suggest that the role of the subject in thinking deserves such critical reflection, and with the wealth of data from cognitive neurosciences, I would like to suggest that "bias" is the important implicit information that has to be addressed in relation to the subject.

With explicit analysis of the relation between bias and subjectivity, we create the opportunity to understand it, predict how it works, and possibly address it when it does not work well, when it comes at an unnecessary cost for us or for others. This kind of analysis can be done for any bias and for any subject. We can try to represent the affects of others in various situations—a process commonly referred to as "empathy" when it occurs implicitly, a process also that usually comes with the unchecked assumption that the representations are accurate, as if we can truly intuit the feelings of others. Some form of quality control of these representations would be advisable, as well as proper reflection on the selectivity of empathy, which can be fairly categorical as we have seen in the fMRI data by Singer et al. (2006).

As we learn more about the relation between bias and subjectivity, and about representing and taking into account the affects of others, we are likely to start thinking differently and more carefully about moral matters—it is perhaps one of the finest features of our species that we are indeed capable of cognitive flexibility in the moral domain, where a number of things of the past are seen today as unthinkably cruel and unfair. Ethics and moral thinking are, and probably should be, continually in flux, as argued most vocally by the contemporary philosopher Peter Singer (2000; not the same Singer as the one from the fMRI study on empathy as a function of perceived fairness). It is, in fact, a logical consequence of the inescapable role of subjectivity in moral decision making—thus, the inevitable impact of historical, geographical, and cultural contingencies.

Rethinking in the ethical domain will usually imply the elucidation of implicit biases, which, once exposed and thought through, might invite a subject to apply a counteractive mechanism. It is a technique employed by the same Peter Singer (2000, p. 5), for instance, when he tries to rethink our eating habits: "Whatever method I employ, I must be aware of the possibility that my own desire to eat meat may lead to bias in my deliberations." In a similar vein, I think cognitive neuroscientists would do well to engage in continual rethinking of experimental practices, for instance, with respect to using monkeys for single-unit recording. There is no question that the technique has produced—and still produces—a wealth of crucial data without which we would know very little indeed about the neural circuits underlying decision making. I have learned immensely from what I have affectionately dubbed "Wurtz-like studies," many of which feature prominently in this book. However, as the cognitive neurosciences evolve, as alternative means become available and improve, with better resolution for human neuroimaging studies and with single-unit studies in rats performing highly sophisticated behavioral tasks, we may wish to rely as little as possible on working with monkeys, that is, with creatures who probably have conscious affects so rich and complex as to deserve being fully represented as those of others—"others" as in "individuals for whom we have an undeniable moral responsibility."

Of course, it is not easy to rethink experimental practices, particularly if they imply a personal investment, say, for researchers who actually work with monkeys. The question is further complicated by a paradox in that the moral maxim of honoring affects as much as we can, recognizing how similar our affects are to those of monkeys, appears to conflict with the sound scientific advice to apply Occam's razor when it comes to interpreting the cognitive abilities of monkeys. Yet, we could counter that Occam's razor by another, pointing to the shared evolutionary background of humans and monkeys and the sound scientific advice to think that "if closely related species act the same, the underlying process probably is the same too." The quoted words are from Frans de Waal (1996, p. 64) on the evolution of moral thinking, including phenomena such as reciprocity and protection of the weaker—all part of the repertoire that De Waal recognized in nonhuman primates.

To be sure, it is not my purpose here to advocate that we should abandon single-unit recording with monkeys. There are certainly types of questions—important questions—that presently can be addressed with this technique and no other. However, I do believe continual rethinking is warranted. Some questions may be less important or less urgent, and in a few cases, it might be possible to obtain equivalent data with rats. Taking the richness and complexity of conscious experience as a criterion, based on knowledge gathered in neuroscience, it would seem fair to think, ethically, that monkeys are less preferable than rats as an animal model. Therefore, if science can progress equally with an experiment in either species, I would propose that we should perform the experiment with rats instead of monkeys.

To this ethical note I will readily add that I do not take my proposal to be neutrally or objectively "a new true object." It is simply one personal view offered for discussion in a public forum, acknowledging that Utopia does not exist and that the world is full of biases, including my own. So if I said "we" a little too frequently, it is perhaps because I have an overeager tendency to hear myself as if you had heard me. I, that is, here with the words of Susan Howe (2007, p. 96):

I heard myself as if you
had heard me utopically
before reflection I heard
you outside only inside
sometimes only a word
So in a particular world
as in the spiritual world

Coda: The Pleasure of Prediction Error

So the music stops. There will be no new poem, no chapter 8—with the formulas of Bayes and bias applied, the various wishes come true, the many fears materialized, the familiarity fallacy committed, the proximity trap fallen into, the game sometimes won and sometimes lost but always played, and always played better as the stakes were higher and the lucky slots fewer, less being more, and biases working with biases, for or against, I think I cannot get any closer to my Utopia of a book with not a species of bias left untouched.

I have endeavored to discuss, to describe, and sometimes literally to draw the underlying neural mechanisms in as much detail as the empirical data can safely afford. It goes without saying that the product of this effort reflects a single-person account of a state of things at a particular moment in time, and so it will inevitably be incomplete and even outdated before the book is actually printed and ready to be read. However, it is only through exercises such as this one that the basic architecture of phenomena, rather than the minute ramifications of a single phenomenon, can be made accessible as an object for revision, reevaluation, and reorientation in research and as an object for communication with people outside the inner circle of researchers in the cognitive neurosciences, from students and aspiring researchers to readers with a primary interest in other domains of human inquiry. If the account proposed in this book has a chance to retain some kind of validity beyond the moment of its composition, it is only to the extent it can be used as a tool in the application, and the further development, of theories, paradigms, and empirical questions relating to decision making and its neural underpinnings.

I, for one, will certainly be happy to see my account of *How Neural Circuits Weigh the Options* be rebutted, further specified, upgraded, or otherwise improved by new ideas and unexpected findings. The prediction error will come with a little pulse of pleasure, not unlike what presumably occurs when dopamine neurons are excited into reporting a positive change to the current mental model of things as they are. The overthrowing of old conceptions, the

Popperian ideal for scientific progress, comes not with melancholy for what is then lost but with a jolt of joy, perhaps characteristic of a particular type of human (also known as "nerd"), hooked on the acquisition of knowledge.

What I Seek in Speech

Concretely, then, I would like to offer my account as a challenge, saying this is my best understanding at the moment and I would really appreciate it if anyone has any comments or suggestions about where I was mistaken and what I have missed, and I would be even happier if somehow those comments or suggestions can find their way back to me. With Jacques Lacan (2004, p. 94), one of the masters of the talking cure:

> What I seek in speech is the response of the other. What constitutes me as subject is my question. In order to be recognized by the other, I utter what was only in view of what will be. In order to find him, I call him by a name that he must assume or refuse in order to reply to me.

True to the spirit of exploration and experiment in science and elsewhere, I think *The Anatomy of Bias* is more of a question than an answer, and perhaps my subtitle should have been *How Do Neural Circuits Weigh the Options?* ending with a question mark. I have, or "what constitutes me as a subject" has, uttered "what was only in view of what will be"—a prospective statement, a predictive image that goes out to look for a return, a confirmation or a rejection, not from within but from without, from you, the reader.

There is something curiously recursive about this process with respect to a monograph on bias, with anticipatory activity on the topic of wishful seeing and other forms of perceptual or cognitive representation of what is not quite there yet. In a roundabout way, it reiterates perhaps the most salient point of this book—that mechanisms of bias are so ingrained in human thought that we encounter them in many different guises in a wide variety of contexts and situations. Bringing the concept of bias to the fore, my purpose has been to promote it explicitly as a core question in any investigation of decision making.

In fact, the concept may well prove useful beyond its traditional playground of decision making, where it was articulated with unsurpassable precision in the framework of signal detection theory as well as that of the LATER model, to name only the examples that I have returned to again and again. The discussion in chapter 5, "The Proximity Trap," on the role of bias in perceptual organization asks for a more thorough analysis in relation to the creation of objects, semantic objects, or things that mean things. Indeed, the very notion of bias and its predictive imagery is intricately linked not only with the subject, the one who shows the bias, but also with the object, perhaps not quite Heidegger's "true new object" (encountered in chapter 7), but in any case an object in the sense of *something*, a thing or an event on which the bias is focused.

For most of the present account, I have simply taken objects as givens—some objects had established positive connotations and acted as rewards, whereas other objects carried known negative implications and invited avoidance behavior. The learning mechanisms I have addressed usually involved nothing more than mere adaptation to the given meanings of objects. I have, by and large, avoided the tricky question of exactly how objects are defined and represented. Yet, the dynamics of object formation, of putting together meaningful units and using labels or words to juggle them, to link them in syntactic ways, all of these things are definitely worth a closer look in relation to bias and neural mechanisms of anticipatory processing that tend to favor some options over other.

Perhaps the objects can be thought of as attractors that define the degrees of freedom for a dynamic set of neural processes on their way in a chaotic itinerancy. Having objects in mind, or thinking things, might correspond to neural processes homing in on a stable state, a balance, a form of harmony in representation, in which the brightest lucidity is reached when objects appear most sharply and singularly. How do the attractors emerge? Are they the projected endpoints of given biases in new contexts? Could the strength of biases toward such projected endpoints reflect the degrees of freedom for the dynamics of neural activity? Could biases facilitate the force of gravity toward these points of final stability by building privileged connections among features in object files? That is, could the detection of one feature evoke a bias to detect a feature associated with the same object? How dynamic or context-dependent are such biases in the making of associations? How does the definition of objects relate to "biological significance," or the basic needs and tendencies of subjects?

The concept of bias generates questions that would not have come to my mind otherwise. Whether the questions lead to an interesting place, or whether they are somehow absurd or nonsensical, I do not know. Not yet, anyway. I will have to try them out in the laboratory, and compare them with findings and ideas communicated by others, and if all works well, I might write another single-person account of a state of things at a particular moment in time, someday in an as yet undefined future. In the meantime, the object of that account is already beginning to take shape in my mind—I can even see the title before me, *The Gravity of Harmony*, and the subtitle, *How Neural Circuits Create Objects of Thought*.

Objects Represented to the Eye

"What good is a mystery?" I asked early on in the course of *The Anatomy of Bias*. Perhaps the answer can be found, again, in the process of wishful seeing, with emphasis on *wishful* rather than on *seeing*. The right kind of scientific "mystery" comes with a sense of wonder and in defiance of its etymology, which is probably rooted in the Greek *myein* ("to close" or "to shut," presumably the eyes or the mouth). I believe it positively motivates and stimulates us not only to search in a particular direction but, more fundamentally, to continue looking in the first place. The creation of concepts, along with definitions of what it is we might wish to see, sets us going along specific lines of investigation, indicating a route as well as actually pushing us forward. It is not merely about sound methodological thinking but also about fully acknowledging the basic joy that underscores the scientific enterprise.

Reversing the arrow in John Keats's most famous line of verse, "A thing of beauty is a joy forever" (here quoted from the heart, but to be found in Book 1 of *Endymion*), I think the boundless joy, the fundamental happiness of being able to work in science and dedicate our lives to the pursuit of knowledge implies that there is more at play than the issue of power. With our concepts and experiments, we are creating things of beauty. With Spinoza, we might even claim that somehow beauty is a natural extension of well-being, so that even dyed-in-the-wool utilitarian scientists, who adamantly claim to concentrate exclusively on manufacturing things that produce some kind of practical benefit, are, whether they like it or not, bringing beauty to the world. Here is a quote from *Ethics* (Spinoza, 2001, p. 40): "[I]f the motion by which the nerves are affected by means of objects represented to the eye conduces to well-being, the objects by which it is caused are called beautiful; while those exciting a contrary motion are called deformed." I like these words not just for the connection they establish between well-being and beauty but also for the curious lapse (due to the translator?) with "represented"—objects *re*-presented to the eye should be subtly different from the physical ones that are initially presented to the eye in the form of refracted rays of light. I am sure I read too much in the quote when I think I detect an oblique reference here to the internal nature of the process, implying that the eye is in the mind and that the objects in question are, in fact, a complex product of real things and what comes before them by the way of projections, wishful images, or hypothetic anticipations, as a function of bias toward well-being.

Rereading texts such as these, even at the risk of misreading, I would like to argue, is just one of the ways in which we can connect with other modes of thought and allow the aesthetic dimension to have its proper place in science as one of three dimensions that are intricately interwoven. Call me naive, but striving toward the convergence of truth, goodness, and beauty is more than merely an idealist's vision of science. I think it is a crucial task for any scientist who wishes, like I do, a central position for science in society. The quest for truth is prominent enough in the minds of most scientists, but too often the aesthetic and moral dimensions are neglected or left to others to work out. This is a problem because these two dimensions are particularly crucial in the exchange between science and other domains of human society.

It is a lack of interest in the aesthetic dimension, or an overemphasis on "rationality," that leaves large sections of society with the erroneous image that science is a cold and boring affair, more of a chore than a pleasure. It is a lack of interest in the moral dimension, or a consistent evasion of difficult ethical questions, that leaves some people in the damaging illusion that scientists have no heart and a skewed understanding of what is really important. We will find more enthusiasm for science, both in terms of broad public support and in numbers of young aspiring researchers, if we explicitly allow goodness and beauty to be essential partners in the scientific enterprise.

This is not to say that everyone should dig up exotic quotes or reason on the basis of personal anecdotes, the way I have done here, but I do wish to offer a plea in favor of a scientific discourse that goes beyond the purely technical analysis of experimental protocols and data, without jumping to the other extreme of popularizing and simplifying to the point of actually missing what was there to learn from the research. Ideally—not a word to be afraid of—the aesthetic and moral dimensions in scientific discourse work to amplify the facts revealed by research. The truths will sound better and more vigorously, and so, if age-old poetic principles count for something, chances increase that the findings will be heard, the knowledge shared.

This Makes the Blood-Vessels Flush

If the convergence of truth, beauty, and goodness sounds impossibly heavenly, your cheeks flushed with vicarious shame for my hopeless gullibility, I should add in would-be mature sobriety that I see the convergence solely as a set of guidelines and orientations for the praxis of science. It is, in the first place, the research conduct to which the evaluative categories apply—a researcher's way of doing science can be called truthful, morally good, and aesthetically pleasing. This, in and of itself, gives no guarantee whatsoever with respect to the quality of the data that are produced in the research. We may have a truthful, morally good, and aesthetically pleasing praxis that fails to break new ground. None of the three evaluative categories provide a metric of genius.

What constitutes genius, or how a researcher might change the shape of a research field, cannot be grasped by considering the soundness of research praxis and instead must be gathered from the actual content of the work. Not *how*, but *what*—a different question altogether, possibly something for *The Gravity of Harmony*.

With the focus on bias in decision making, the present monograph naturally brings the question of research praxis to the foreground. The three evaluative categories can each be seen as polarized dimensions, where the guidelines for research imply a set of biases, preferring truth over falsehood, good over evil, beauty over ugliness. Bringing these biases to the surface, to our conscious awareness, allows us to apply them more carefully, more systematically, and more successfully. This counsel represents the final raison d'être of *The Anatomy of Bias*—the core proposal is that grasping our tendencies, leanings, and prejudices, understanding how they work, realizing which are too strong or too weak, will give us an opportunity to improve our decision making. Above all, I have aimed to emphasize the role of bias as a mechanism that influences how decisions are made, any type of decision.

This role of bias has, I think, remained too covert in much contemporary research on decision making and perceptual processing, because of what I see as the improper usage of the beautiful but ill-defined concept of attention—beautiful in the language of William James, and the way in which it seems to give access to the more complex and fascinating aspects of human experience, but ill-defined as researchers have variously used it to imply either clarity of vision and improvement of information processing (what I refer to as "increased sensitivity") or, more generally, prioritization and selection of information (with the application of bias and/or sensitivity). Blurring between these two meanings of attention is particularly harmful when researchers design experiments that test for prioritization and selection—opening the door to a mixture of bias and sensitivity effects—and then discuss the results wearing blinkers, using only the more specific concept of attention, referring to increased sensitivity. In this way, effects of bias are sometimes misattributed to sensitivity.

We would do well to concentrate more carefully on the usage of concepts as tools for research, as essential components of sound methodology. It asks for a continued commitment to analytic and creative thinking in the development of language, not merely the "scientific language" of research reports, which tends to be less scientific in introductions and discussions, and occasionally downright nonscientific in titles. The analytic and creative thinking is needed also with respect to the language we use when we communicate about the research to wider audiences. Several philosophers have claimed entitlement to analytic thinking, and many poets are notoriously confident about believing the power of creation to be theirs, but I would rather incorporate the special concern for language as an intrinsic element of the scientific enterprise (in fact, I think the integration would naturally happen when we give more weight to the aesthetic and moral dimensions in research). The categories of "philosopher," "scientist," and "poet" are probably slightly suspicious anyway, curiously restrictive as they are, and restrictive in different ways as a function of who uses these labels.

Perhaps I should give an example of the kind of special concern for language I would like to see practiced more often in the course of human inquiry. My favorite prototype of the philosopher–scientist–poet model would have to be Charles Sherrington, arguably the first truly systems-oriented neuroscientist, who among many other things may be remembered for his creativity in coming up with a word like *synapse* (from the Greek *syn-*, "together," and *haptein*, "to clasp"). Sherrington published three major works: one unmistakably scientific in 1906, *The Integrative Action of the Nervous System* (Sherrington, 1961), in which he related the concept of reflexes to purposive action; one combatively poetic in 1925, *The Assaying of Brabantius and Other Verse* (cited in Fuller, 2007), in which he indulged in the most eccentric mannerisms; and one primarily philosophical in 1940, *Man on His Nature* (Sherrington, 1955), in which he explored the continued relevance of the writings of Jean Fernel (a sixteenth-century French physician, Cartesian before Descartes, and, being the inventor of the concept, the first true physiologist). It is from *Man on His Nature* that I would like to sample a few excerpts, for no other reason than to marvel at the language used in scientific communication.

Though Sherrington fully achieved a convergence of the true, the good, and the beautiful, this does not mean that he fell prey to an unchecked optimism bias. A particularly delightful *and* deadly serious passage is where he tackles "a gnat, called anopheles, from the Greek word for 'hurtful'" (Sherrington, 1955, p. 274), as a challenge to any creationist vision of nature (see also figure C.1, a schematic representation of unusual aesthetic quality—note especially the wonderful waviness of the words embedded in the drawing). The topic of mosquitoes, parasites, and malaria moved him so much that he (or whoever transposed his text on paper) started stuttering, "This is a a grievous disease" (sic, p. 276). Here is how Sherrington describes with perfect clarity, not a word too much, the mosquito's action on human skin (p. 274):

The so-called "bite" happens like this. The gnat when she alights on the skin tests the place with her labellum. Then steadying her head against the skin she stabs by styles with dagger points and saw-like edges. Swaying her head as she uses her mouth-parts she works these through the skin. They go through and among the blood-vessels, and

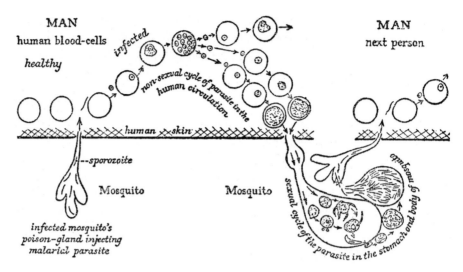

Figure C.1
Exchange of individual (parasitic) information among humans via a biological carrier (a mosquito), or a schematic presentation of the processes underlying malaria, borrowed from Charles Sherrington's (1955) classic *Man on His Nature*.

carry with them a tiny tube like a hollow needle, close behind the stabbing style. It leads from a poison-gland. It injects a droplet of juice into the stabbed wound. This makes the blood-vessels flush; they bring more blood to the stabbed spot. Also the juice delays the clotting of the blood which might baulk the gnat of her full meal, by cutting short the yield of blood from the tiny wound.

The little narrative is both extremely effective in conveying a lot of detailed scientific information and rich enough in imagery and common language to be accessible to just about any reader. If this is popularizing science, it achieves its purpose without any distraction from the complexity of the topic. Sherrington tops it off with little observations that connect with a more contemplative (aesthetic) mode of thought: "Nature has provided her with special tools and a special zest for thoroughness" and "[r]ich food it is, human blood" (both still on p. 274). I feel the blood-vessels flush in my cheeks when I read a passage like this. Is it out of shame? Or jealousy? I wish we could all write with the same special zest. I think it is well within the interest of science if at least we try.

The Endeavor Has Just Begun

With the notes on the interest of science it should be clear that much of what I have written and discussed in the present monograph is prospective in spirit. Focusing on the usefulness of particular concepts and the soundness of various methodological approaches, the underlying assumption has been that most of the scientific work lies ahead of us. Now at the end of my journey into *The Anatomy of Bias*, I feel compelled to quote the words of my teacher, Okihide Hikosaka, some ten years ago, concluding a brief opinion piece: "The endeavor has just begun, disentangling the unpredictable mind and behavior" (Hikosaka, 1999, p. 338). The words came in response to a query from *Brain Research Bulletin* to highlight a truly groundbreaking research effort in neuroscience (the honor went, not surprisingly and quite justifiably, to Wurtz and Goldberg for "probing consciousness with an electrode"). The words of Wurtz's finest disciple were straightforward and, again, entirely prospective. Whether mind and behavior are simply unpredictable, I am not sure—maybe we can think in degrees of predictability. Also the idea that mind and behavior should be *disentangled* could spark off another debate. But that the endeavor has only just begun, I could not agree more.

In the meantime, though, I find myself having to put a stop somewhere in the present tense of the endeavor for the purpose of completing something that can be committed to the press. It will be a snapshot then, taken deep in December of the year 2008. I look forward to much exciting new research in 2009 and beyond, but for now I think I am already in the (antipodean?) "Fields of Summer," as composed by the great Chuvash poet Gennady Aygi for *Child-and-Rose* (2003, p. 132):

and concluding—so we too must fall silent:
"far off"—it is only a quiet sound
of a distant homeland...(oh yes: like a hymn...)
and this
is straightway
long ago

Bibliography

Alighieri, D. (1995). *The Divine Comedy: Inferno; Purgatorio; Paradiso*. (A. Mandelbaum, Transl.) New York, NY: Alfred A. Knopf (Everyman's Library). (Original work written between 1308 and 1321.)

Allport, D. A. (1989). Visual attention. In: M. I. Posner (Ed.), *Foundations of Cognitive Science* (pp. 631–682). Cambridge, MA: MIT Press.

Amari, S., Park, H., & Ozeki, T. (2006). Singularities affect dynamics of learning in neuromanifolds. *Neural Computation, 18,* 1007–1065.

Andersen, A. K., Christoff, K., Stappen, I., Panitz, D., Ghahremani, D. G., Glover, G., Gabrieli, J. D., & Sobel, N. (2003). Dissociated neural representations of intensity and valence in human olfaction. *Nature Neuroscience, 6,* 196–202.

Anderson, J. R. (1983). A spreading activation theory of memory. *Journal of Verbal Learning and Verbal Behavior, 22,* 261–295.

Andretic, R., Van Swinderen, B., & Greenspan, R. J. (2005). Dopaminergic modulation of arousal in *Drosophila*. *Current Biology, 15,* 1165–1175.

Anonymous. (1736). *An Introduction to the Doctrine of Fluxions, and a Defence of the Mathematicians against the Objections of the Author of The Analyst*. London, U.K.: John Noon. (Attributed to Thomas Bayes.)

Anonymous. (2005). *The Bible, King James Version, Book 20: Proverbs*. Project Gutenberg, EText-No. 8020 [http://www.gutenberg.org]. (Accessed July 10, 2008.) (Translation first published 1611.)

Antes, J. R. (1974). The time course of picture viewing. *Journal of Experimental Psychology, 103,* 62–70.

Aygi, G. (2003). *Child-and-Rose*. (P. France, Transl.) New York, NY: New Directions Books.

Bacon, F. (2000). *The New Organum*. (L. Jardine & M. Silverthorne, Ed.; M. Silverthorne, Transl.) Cambridge, U.K.: Cambridge University Press. (Original work published 1620.)

Badiou, A. (2005). *Infinite Thought*. (O. Feltham & J. Clemens, Transl.). London, U.K.: Continuum. (Translation first published 2003.)

Badiou, A. (2007). *Being and Event*. (O. Feltham, Transl.). London, U.K.: Continuum. (Original work published 1988; translation first published 2005.)

Band, G. P., Van der Molen, M. W., & Logan, G. D. (2003). Horse-race model simulations of the stop-signal procedure. *Acta Psychologica, 112,* 105–142.

Barthes, R. (2000). *Camera Lucida: Reflections on Photography*. (R. Howard, Transl.) New York, NY: Hill and Wang. (Original work published 1980; translation first published 1981.)

Basso, M. A., & Wurtz, R. H. (1997). Modulation of neuronal activity by target uncertainty. *Nature, 389,* 66–69.

Baum, W. M. (1974). On two types of deviation from the matching law: Bias and undermatching. *Journal of the Experimental Analysis of Behavior, 22,* 231–242.

Bayes, T. (1763). An essay towards solving a problem in the doctrine of chances. (R. Price, Ed.) *Philosophical Transactions of the Royal Society of London, 53,* 370–418. [http://www.stat.ucla.edu/history/essay.pdf]. (Accessed July 16, 2008; pages numbered 1–23.)

Beauregard, M. (2007). Mind does matter: Evidence from neuroimaging studies of emotional self-regulation, psychotherapy, and placebo effect. *Progress in Neurobiology, 81,* 218–236.

Bechara, A., Damasio, H., Tranel, D., & Damasio, A. R. (1997). Deciding advantageously before knowing the advantageous strategy. *Science, 275,* 1293–1295.

Beck, A. T. (1987). Cognitive model of depression. *Journal of Cognitive Psychotherapy, 1,* 2–27.

Belova, M. A., Paton, J. J., Morrison, S. E., & Salzmann, C. D. (2007). Expectation modulates neural responses to pleasant and aversive stimuli in primate amygdala. *Neuron, 55,* 970–984.

Benjamin, W. (1968). *Illuminations.* (H. Zohn, Transl.) New York, NY: Harcourt, Brace & World.

Bennett, M. R., & Hacker, P. M. S. (2003). *Philosophical Foundations of Neuroscience.* Malden, MA: Blackwell Publishing.

Bhatia, K. P., & Marsden, C. D. (1994). The behavioural and motor consequences of focal lesions of the basal ganglia in man. *Brain, 117,* 859–876.

Biederman, I. (1987). Recognition-by-components: A theory of human image understanding. *Psychological Review, 94,* 115–147.

Biederman, I., & Gerhardstein, P. C. (1993). Recognizing depth-rotated objects: Evidence and conditions for three-dimensional viewpoint invariance. *Journal of Experimental Psychology: Human Perception and Performance, 19,* 1162–1182.

Biederman, I., Mezzanotte, R. J., & Rabinowitz, J. C. (1982). Scene perception: Detecting and judging objects undergoing relational violations. *Cognitive Psychology, 14,* 143–177.

Bisley, J. W., & Goldberg, M. E. (2003). Neuronal activity in the lateral intraparietal area and spatial attention. *Science, 299,* 81–86.

Bjork, J. M., & Hommer, D. W. (2007). Anticipating instrumentally obtained and passively-received rewards: A factorial fMRI investigation. *Behavioural Brain Research, 177,* 165–170.

Blake, W. (1975). *The Marriage of Heaven and Hell.* (G. Keynes, Ed.) Oxford, U.K.: Oxford University Press. (Original work completed 1790.)

Blanchot, M. (1989). *The Space of Literature.* (A. Smock, Transl.) Lincoln, NE: University of Nebraska Press. (Original work published 1955.)

Blank, H., Nestler, S., von Collani, G., & Fischer, V. (2008). How many hindsight biases are there? *Cognition, 106,* 1408–1440.

Blanton, H., & Jaccard, J. (2006). Tests of multiplicative models in psychology: A case study using the unified theory of implicit attitudes, stereotypes, self-esteem, and self-concept. *Psychological Review, 113,* 155–169.

Boden, M. A. (2008). *Mind as Machine: A History of Cognitive Science.* (Two volumes.) Oxford, U.K.: Oxford University Press. (Original work published 2006.)

Bogacz, R., Brown, E., Moehlis, J., Holmes, P., & Cohen, J. D. (2006). The physics of optimal decision making: A formal analysis of models of performance in two-alternative forced-choice tasks. *Psychological Review, 113,* 700–765.

Borg, J. S., Lieberman, D., & Kiehl, K. A. (2008). Infection, incest, and iniquity: Investigating the neural correlates of disgust and morality. *Journal of Cognitive Neuroscience, 20,* 1529–1546.

Borges, J. L. (1999). *Selected Nonfictions.* (E. Weinberger, Ed.; E. Allen, S. J. Levine, & E. Weinberger, Transl.) New York, NY: Viking Penguin.

Botvinick, M. M., Braver, T. S., Barch, D. M., Carter, C. S., & Cohen, J. D. (2001). Conflict monitoring and cognitive control. *Psychological Review, 108,* 624–652.

Boucher, L., Palmeri, T. J., Logan, G. D., & Schall, J. D. (2007). Inhibitory control in mind and brain: An interactive race model of countermanding saccades. *Psychological Review, 114,* 376–397.

Bowlby, J. (1973). *Attachment and Loss (vol. 2). Separation: Anxiety and Anger.* London, U.K.: Hogarth Press.

Bowlby, J. (1980). *Attachment and Loss (vol. 3). Loss: Sadness and Depression.* London, U.K.: Hogarth Press.

Bowlby, J. (1990). *A Secure Base: Parent–Child Attachment and Healthy Human Development.* New York, NY: Basic Books. (Original work published 1988.)

Bowlby, J. (1999). *Attachment and Loss (vol. 1). Attachment.* New York, NY: Basic Books. (Original work published 1969.)

Boyd, R., Gintis, H., Bowles, S., & Richerson, P. J. (2003). The evolution of altruistic punishment. *Proceedings of the National Academy of Sciences of the United States of America, 100,* 3531–3535.

Breuer, J., & Freud, S. (2000). *Studies on Hysteria.* (J. Strachey, Transl.) New York, NY: Basic Books. (Original work published 1895; translation first published 1957.)

Britten, K. H., Shadlen, M. N., Newsome, W. T., & Movshon, J. A. (1992). The analysis of visual motion: A comparison of neuronal and psychophysical performance. *Journal of Neuroscience, 12,* 4745–4765.

Broadbent, D. E. (1961). *Perception and Communication.* London, U.K.: The Scientific Book Guild. (Original work published 1958.)

Brosnan, S. F., & De Waal, F. B. M. (2003). Monkeys reject unequal pay. *Nature, 425,* 297–299.

Brown, J. W., & Braver, T. S. (2005). Learned predictions of error likelihood in the anterior cingulate cortex. *Science, 307,* 1118–1121.

Brown, P., Chen, C. C., Wang, S., Kühn, A. A., Doyle, L., Yarrow, K., Nuttin, B., Stein, J., & Aziz, T. (2006). Involvement of human basal ganglia in offline feedback control of voluntary movement. *Current Biology, 16,* 2129–2134.

Buckholtz, J. W., Asplund, C. L., Dux, P. E., Zald, D. H., Gore, J. C., Jones, O. D., & Marois, R. (2008). The neural correlates of third-party punishment. *Neuron, 60,* 930–940.

Burke, K. A., Franz, T. M., Miller, D. N., & Schoenbaum, G. (2008). The role of the orbitofrontal cortex in the pursuit of happiness and more specific rewards. *Nature, 454,* 340–344.

Burton, R. (2001). *The Anatomy of Melancholy.* (H. Jackson, Ed.). New York, NY: The New York Review of Books. (Original work published 1621.)

Buschman, T. J., & Miller, E. K. (2007). Top–down versus bottom–up control of attention in the prefrontal and posterior parietal cortices. *Science, 318,* 1860–1862.

Calvin, W. H. (1996). *The Cerebral Code: Thinking a Thought in the Mosaics of the Mind.* Cambridge, MA: MIT Press.

Camerer, C. F. (2003). *Behavioral Game Theory: Experiments in Strategic Interaction.* Princeton, NJ: Princeton University Press.

Caplan, L. R., Schmahmann, J. D., Kase, C. S., Feldmann, E., Baquis, G., Greenberg, J. P., Gorelick, P. B., Helgason, C., & Hier, D. B. (1990). Caudate lesions. *Archives of Neurology, 47,* 133–143.

Carpenter, R. H. S. (1981). Oculomotor procrastination. In: D. F. Fisher, R. A. Monty, & J. W. Senders (Eds.), *Eye Movements: Cognition and Visual Perception* (pp. 237–246). Hillsdale, NJ: Lawrence Erlbaum.

Carpenter, R. H. S. (1999). A neural mechanism that randomises behaviour. *Journal of Consciousness Studies, 6,* 13–22.

Carpenter, R. H. S. (2004). Contrast, probability, and saccadic latency: Evidence for independence of detection and decision. *Current Biology, 14,* 1576–1580.

Carpenter, R. H. S., & Williams, M. L. L. (1995). Neural computation of log likelihood in control of saccadic eye movements. *Nature, 377,* 59–62.

Carr, T. H. (1992). Automaticity and cognitive anatomy: Is word recognition "automatic"? *American Journal of Psychology, 105,* 201–237.

Carson, A. (2003). *If Not, Winter: Fragments of Sappho.* New York, NY: Vintage Books. (Translation first published 2002.)

Celan, P. (2001). *Selected Poems and Prose of Paul Celan.* (J. Felstiner, Transl.) New York, NY: W.W. Norton. (Translation first published 2000.)

Chomsky, N. (2002). *Understanding Power: The Indispensable Chomsky.* (P. R. Mitchell & J. Schoeffel, Eds.) New York, NY: The New Press.

Christoph, G. R., Leonzio, R. J., & Wilcox, K. S. (1986). Stimulation of the lateral habenula inhibits dopamine-containing neurons in the substantia nigra and ventral tegmental area of the rat. *Journal of Neuroscience, 6,* 613–619.

Churchland, A. K., Kiani, R., & Shadlen, M. N. (2008). Decision-making with multiple alternatives. *Nature Neuroscience, 11,* 693–702.

Coe, B., Tomihara, K., Matsuzawa, M., & Hikosaka, O. (2002). Visual and anticipatory bias in three cortical eye fields of the monkey during an adaptive decision-making task. *Journal of Neuroscience, 22,* 5081–5090.

Cohen, J. D., & Servan-Schreiber, D. (1992). Context, cortex, and dopamine: A connectionist approach to behavior and biology in schizophrenia. *Psychological Review, 99,* 45–77.

Collins, A. M., & Loftus, E. F. (1975). A spreading-activation theory of semantic processing. *Psychological Review, 82,* 407–428.

Cregg, B. (2003). *The First Scientist: A Life of Roger Bacon.* Cambridge, MA: Da Capo Press.

Critchley, H. D., Mathias, C. J., & Dolan, R. J. (2001). Neural activity in the human brain relating to uncertainty and arousal during anticipation. *Neuron, 29,* 537–545.

Cunningham, W. A., Johnson, M. K., Raye, C. L., Gatenby, J. C., Gore, J. C., & Banaji, M. R. (2004a). Separable neural components in the processing of black and white faces. *Psychological Science, 15,* 806–813.

Cunningham, W. A., Raye, C. L., & Johnson, M. K. (2004b). Implicit and explicit evaluation: fMRI correlates of valence, emotional intensity, and control in the processing of attitudes. *Journal of Cognitive Neuroscience, 16,* 1717–1729.

Damasio, A. R. (1999). *The Feeling of What Happens: Body and Emotion in the Making of Consciousness.* New York, NY: Harcourt.

Damasio, A. R. (2003). *Looking for Spinoza: Joy, Sorrow, and the Feeling Brain.* New York, NY: Harcourt.

Damasio, A. R. (2005). *Descartes' Error: Emotion, Reason, and the Human Brain.* New York, NY: Penguin. (Original work published 1994.)

Damasio, A. R., Tranel, D., & Damasio, H. (1991). Somatic markers and the guidance of behavior: Theory and preliminary testing. In: H. S. Levin, H. M. Eisenberg, & A. L. Benton (Eds.), *Frontal Lobe Function and Dysfunction* (pp. 217–229). Oxford, U.K.: Oxford University Press.

D'Ardenne, K., McClure, S. M., Nystrom, L. E., & Cohen, J. D. (2008). BOLD responses reflecting dopaminergic signals in the human ventral tegmental area. *Science, 319,* 1264–1267.

Darwin, C. (1965). *The Expression of the Emotions in Man and Animals.* Chicago, IL: The University of Chicago Press. (Original work published 1872.)

Daw, N. D., Niv, Y., & Dayan, P. (2005). Uncertainty-based competition between prefrontal and dorsolateral striatal systems for behavioral control. *Nature Neuroscience, 8,* 1704–1711.

Dawkins, R. (2006). *The Selfish Gene: 30th Anniversary Edition—With a New Introduction by the Author.* Oxford, U.K.: Oxford University Press. (Original work published 1976.)

Deacon, T. W. (1997). *The Symbolic Species: The Co-evolution of Language and the Brain.* New York, NY: W.W. Norton.

De Fockert, J. W., Rees, G., Frith, C. D., & Lavie, N. (2001). The role of working memory in visual selective attention. *Science, 291,* 1803–1806.

De Graef, P., Christiaens, D., & d'Ydewalle, G. (1990). Perceptual effects of scene context on object identification. *Psychological Research, 52,* 317–329.

DeLanda, M. (2004). *Intensive Science and Virtual Philosophy.* New York, NY: Continuum. (Original work published 2002.)

Deleuze, G. (1994). *Difference and Repetition.* (P. Patton, Transl.) New York, NY: Columbia University Press.

Deleuze, G., & Guattari, F. (1994). *What Is Philosophy?* (H. Tomlinson & G. Burchell, Transl.) New York, NY: Columbia University Press. (Original work published 1991.)

Delgado, M. R., Frank, R. H., & Phelps, E. A. (2005). Perceptions of moral character modulate the neural systems of reward during the trust game. *Nature Neuroscience, 8,* 1611–1618.

Delgado, M. R., Nearing, K. I., LeDoux, J. E., & Phelps, E. A. (2008a). Neural circuitry underlying the regulation of conditioned fear and its relation to extinction. *Neuron, 59,* 829–838.

Delgado, M. R., Nystrom, L. E., Fissell, C., Noll, D. C., & Fiez, J. A. (2000). Tracking the hemodynamic responses to reward and punishment in the striatum. *Journal of Neurophysiology, 84,* 3072–3077.

Delgado, M. R., Schotter, A., Ozbay, E. Y., & Phelps, E. A. (2008b). Understanding overbidding: Using the neural circuitry of reward to design economic auctions. *Science, 321,* 1849–1852.

DeLillo, D. (1997). *Underground.* New York, NY: Simon & Schuster.

De Martino, B., Kumaran, D., Seymour, B., & Dolan, R. J. (2006). Frames, biases, and rational decision-making in the human brain. *Science, 313,* 684–687.

Denève, S., Duhamel, J. R., & Pouget, A. (2007). Optimal sensorimotor integration in recurrent cortical networks: A neural implementation of Kalman filters. *Journal of Neuroscience, 27,* 5744–5756.

Dennett, D. C. (1989). *The Intentional Stance.* Cambridge, MA: MIT Press. (Original work published 1987.)

Dennett, D. C. (1991). *Consciousness Explained.* Boston, MA: Little, Brown and Company.

Dennett, D. C. (1996). *Darwin's Dangerous Idea: Evolution and the Meanings of Life.* New York, NY: Touchstone. (Original work published 1995.)

De Quervain, D. J. F., Fischbacher, U., Treyer, V., Schellhammer, M., Schnyder, U., Buck, A., & Fehr, E. (2004). The neural basis of altruistic punishment. *Science, 305,* 1254–1258.

Descartes, R. (1993). *Meditations on First Philosophy.* (D. A. Cress, Transl.) Indianapolis, IN: Hacket Publishing. (Original work published 1641.)

Descartes, R. (2007). *Discourse on Method.* (P. Kraus & F. Hunt, Eds.; R. Kennington, Transl.) Newburyport, MA: Focus Publishing. (Original work published 1637.)

Desimone, R., & Duncan, J. (1995). Neural mechanisms of visual selective attention. *Annual Review of Neuroscience, 18,* 193–222.

Deutsch, J. A., & Deutsch, D. (1963). Attention: Some theoretical considerations. *Psychological Review, 70,* 80–90.

De Waal, F. (1996). *Good Natured: The Origins of Right and Wrong in Humans and Other Animals.* Cambridge, MA: Harvard University Press.

Dickinson, A. (1980). *Contemporary Animal Learning Theory.* Cambridge, U.K.: Cambridge University Press.

Dickinson, A., & Balleine, B. (1994). Motivational control of goal-directed action. *Animal Learning & Behavior, 22,* 1–18.

Dickinson, E. (1961). *The Complete Poems of Emily Dickinson.* (T. H. Johnson, Ed.) New York, NY: Little, Brown and Company.

Ding, L., & Hikosaka, O. (2006). Comparison of reward modulation in the frontal eye field and caudate of the macaque. *Journal of Neuroscience, 26,* 6695–6703.

Dommett, E., Coizet, V., Blaha, C. D., Martindale, J., Lefebvre, V., Walton, N., Mayhew, J. E. W., Overton, P. G., & Redgrave, P. (2005). How visual stimuli activate dopaminergic neurons at short latency. *Science, 307,* 1476–1479.

Donders, F. C. (1869). Over de snelheid van psychische processen. *Nederlands Archief voor Genees- en Natuurkunde, 4,* 117–145.

Donders, F. C. (1969). On the speed of mental processes. (W. G. Koster, Transl.) *Acta Psychologica, 30,* 412–431.

Dorris, M. C., & Munoz, D. P. (1998). Saccadic probability influences motor preparation signals and time to saccadic initiation. *Journal of Neuroscience, 18,* 7015–7026.

Dorris, M. C., Paré, M., & Munoz, D. P. (2000). Immediate neural plasticity shapes motor performance. *Journal of Neuroscience, 20, RC52,* 1–5.

Downing, C. J. (1988). Expectancy and visual–spatial attention: Effects on perceptual quality. *Journal of Experimental Psychology: Human Perception and Performance, 14,* 188–202.

Doya, K., & Ishii, S. (2007). A probability primer. In: K. Doya, S. Ishii, A. Pouget, & R. P. N. Rao (Eds.), *Bayesian Brain: Probabilistic Approaches to Neural Coding* (pp. 3–13). Cambridge, MA: MIT Press.

Doya, K., Ishii, S., Pouget, A., & Rao, R. P. N. (Eds.) (2007). *Bayesian Brain: Probabilistic Approaches to Neural Coding.* Cambridge, MA: MIT Press.

Dragoi, V., Sharma, J., Miller, E. K., & Sur, M. (2002). Dynamics of neuronal sensitivity in visual cortex and local feature discrimination. *Nature Neuroscience, 5,* 883–891.

Dragomoshchenko, A. (1993). From *Phosphor.* (L. Hejinian & E. Balashova, Transl.) *Postmodern Culture, 3* (issue number 2, electronic source, no page numbers).

Duhamel, J. R., Colby, C. L., & Goldberg, M. E. (1992). The updating of the representation of visual space in parietal cortex by intended eye movements. *Science, 255,* 90–92.

Duhem, P. M. M. (1991). *The Aim and Structure of Physical Theory.* (P. P. Wiener, Transl.) Princeton, NJ: Princeton University Press. (Original work published 1906; translation first published 1954.)

Dujardin, F. (1850). Mémoire sur le système nerveux des insectes. *Annales des Sciences Naturelles (Zoologie), 14,* 195–206.

Edelman, G. M. (1987). *Neural Darwinism: The Theory of Neuronal Group Selection.* New York, NY: Basic Books.

Edelman, G. M. (1992). *Bright Air, Brilliant Fire: On the Matter of the Mind.* New York, NY: Basic Books.

Egner, T., Monti, J. M. P., Trittschuh, E. H., Wieneke, C. A., Hirsch, J., & Mesulam, M.-M. (2008). Neural integration of top–down spatial and feature-based information in visual search. *Journal of Neuroscience, 28,* 6141–6151.

Elliott, R., Agnew, Z., & Deakin, J. F. (2008). Medial orbitofrontal cortex codes relative rather than absolute value of financial rewards in humans. *European Journal of Neuroscience, 27,* 2213–2218.

Elliott, R., Friston, K. J., & Dolan, R. J. (2000). Dissociable neural responses in human reward systems. *Journal of Neuroscience, 20,* 6159–6165.

Entus, A., & Bindra, D. (1970). Common features of the "repetition" and "same–different" effects in reaction-time experiments. *Perception & Psychophysics, 7,* 143–148.

Eriksen, C. W., & St. James, J. D. (1986). Visual attention within and around the field of focal attention: A zoom lens model. *Perception & Psychophysics, 40,* 225–240.

Evarts, E. V. (1966). Pyramidal tract activity associated with a conditioned hand movement in the monkey. *Journal of Neurophysiology, 29,* 1011–1027.

Evarts, E. V. (1968). A technique for recording activity of subcortical neurons in moving animals. *Electroencephalography and Clinical Neurophysiology, 24,* 83–86.

Faverey, H. (2004). *Against the Forgetting: Selected Poems*. (F. R. Jones, Transl.). New York, NY: New Directions Books.

Fehr, E., & Fischbacher, U. (2003). The nature of human altruism. *Nature, 425,* 785–791.

Fehr, E., & Schmidt, K. M. (1999). A theory of fairness, competition, and cooperation. *Quarterly Journal of Economics, 114,* 817–868.

Felleman, D. J., & Van Essen, D. C. (1991). Distributed hierarchical processing in the primate cerebral cortex. *Cerebral Cortex, 1,* 1–47.

Fias, W., Lauwereyns, J., & Lammertyn, J. (2001). Irrelevant digits affect feature-based attention depending on the overlap of neural circuits. *Cognitive Brain Research, 12,* 415–423.

Fiorillo, C. D., Newsome, W. T., & Schultz, W. (2008). The temporal precision of reward prediction in dopamine neurons. *Nature Neuroscience, 11,* 966–973.

Fitch, W. T., & Hauser, M. D. (2004). Computational constraints on syntactic processing in a nonhuman primate. *Science, 303,* 377–380.

Fitzsimmons, T., & Gozo, Y. (Eds.) (1993). *The New Poetry of Japan—The 70s and 80s*. Santa Fe, NM: Katydid Press.

Fodor, J. (2000). *The Mind Doesn't Work That Way: The Scope and Limits of Computational Psychology*. Cambridge, MA: MIT Press.

Foucault, M. (1994). *The Order of Things: An Archaeology of the Human Sciences*. (Anonymous, Transl.) New York, NY: Vintage Books. (Original work published 1966; translation first published 1971.)

Freeman, W. J. (2008). A pseudo-equilibrium thermodynamic model of information processing in nonlinear brain dynamics. *Neural Networks, 21,* 257–265.

Freud, S. (1990). *Beyond the Pleasure Principle*. (J. Strachey, Transl.) New York, NY: W.W. Norton. (Original work published 1920; translation first published 1961.)

Fries, P., Reynolds, J. H., Rorie, A. E., & Desimone, R. (2001). Modulation of oscillatory neuronal synchronization by selective visual attention. *Science, 291,* 1560–1563.

Fuller, J. (2007). The poetry of Charles Sherrington. *Brain, 130,* 1981–1983.

Funahashi, S., Bruce, C. J., & Goldman-Rakic, P. S. (1989). Mnemonic coding of visual space in the monkey's dorsolateral prefrontal cortex. *Journal of Neurophysiology, 61,* 331–349.

Gallavotti, G. (1995). Ergodicity, ensembles, irreversibility in Boltzmann and beyond. *Journal of Statistical Physics, 78,* 1571–1589.

Garoff-Eaton, R. J., Kensinger, E. A., & Schacter, D. L. (2007). The neural correlates of conceptual and perceptual false recognition. *Learning & Memory, 14,* 684–692.

Gauthier, I., Hayward, W. G., Tarr, M. J., Anderson, A. W., Skudlarski, P., & Gore, J. C. (2002). BOLD activity during mental rotation and viewpoint-dependent object recognition. *Neuron, 34,* 161–171.

Gauthier, I., Tarr, M. J., Anderson, A. W., Skudlarski, P., & Gore, J. C. (1999). Activation of the middle fusiform "face area" increases with expertise in recognizing novel objects. *Nature Neuroscience, 2,* 568–573.

Gauthier, I., Williams, P., Tarr, M. J., & Tanaka, J. (1998). Training "greeble" experts: A framework for studying expert object recognition processes. *Vision Research, 38,* 2401–2428.

Gibson, J. J. (1950). *The Perception of the Visual World*. Cambridge, MA: Riverside Press.

Gigerenzer, G. (2002). *Reckoning with Risk: Learning to Live with Uncertainty*. London, U.K.: Allen Lane, The Penguin Press.

Glimcher, P. W. (2003). *Decisions, Uncertainty, and the Brain: The Science of Neuroeconomics*. Cambridge, MA: MIT Press.

Gold, J. I., & Shadlen, M. N. (2001). Neural computations that underlie decisions about sensory stimuli. *Trends in Cognitive Sciences, 5,* 10–16.

Goldberg, M. E., & Wurtz, R. H. (1972). Activity of superior colliculus in behaving monkey. II. Effect of attention on neuronal responses. *Journal of Neurophysiology, 35,* 560–574.

Gonsalves, B., Reber, P. J., Gitelman, D. R., Parrish, T. B., Mesulam, M. M., & Paller, K. A. (2004). Neural evidence that vivid imagining can lead to false remembering. *Psychological Science, 15,* 655–660.

Green, D. M. (1964). Consistency of auditory detection judgments. *Psychological Review, 71,* 392–407.

Green, D. M., & Swets, J. A. (1988). *Signal Detection Theory and Psychophysics.* Los Altos, CA: Peninsula Publishing. (Original work published 1966.)

Greene, J. D., Sommerville, R. B., Nystrom, L. E., Darley, J. M., & Cohen, J. D. (2001). An fMRI investigation of emotional engagement in moral judgment. *Science, 293,* 2105–2108.

Greenwald, A. G., Banaji, M. R., Rudman, L. A., Farnham, S. D., Nosek, B. A., & Mellot, D. S. (2002). A unified theory of implicit attitudes, stereotypes, self-esteem, and self-concept. *Psychological Review, 109,* 3–25.

Greenwald, A. G., McGhee, D. E., & Schwartz, J. L. (1998). Measuring individual differences in implicit cognition: The Implicit Association Test. *Journal of Personality and Social Psychology, 74,* 1464–1480.

Grill-Spector, K., Henson, R., & Martin, A. (2006). Repetition and the brain: Neural models of stimulus-specific effects. *Trends in Cognitive Sciences, 10,* 14–23.

Gross, C. G. (1998). *Brain, Vision, Memory: Tales in the History of Neuroscience.* Cambridge, MA: MIT Press.

Groves, P. M., Linder, J. C., & Young, S. J. (1994). 5-hydroxydopamine-labeled dopaminergic axons: Three dimensional reconstructions of axons, synapses and postsynaptic targets in rat neostriatum. *Neuroscience, 58,* 593–604.

Hamilton, W. D. (1975). Innate social aptitudes of man. In: R. Fox (Ed.), *Biosocial Anthropology* (pp. 133–153). London, U.K.: Malaby Press.

Hanes, D. P., & Schall, J. D. (1996). Neural control of voluntary eye movement. *Science, 274,* 427–430.

Harbaugh, W. T., Mayr, U., & Burghardt, D. R. (2007). Neural responses to taxation and voluntary giving reveal motives for charitable donations. *Science, 316,* 1622–1625.

Hardcastle, V. G., & Stewart, C. M. (2003). Neuroscience and the art of single cell recordings. *Biology and Philosophy, 18,* 195–208.

Hare, T. A., O'Doherty, J., Camerer, C. F., Schultz, W., & Rangel, A. (2008). Dissociating the role of the orbitofrontal cortex and the striatum in the computation of goal values and prediction errors. *Journal of Neuroscience, 28,* 5623–5630.

Haruno, M., Kuroda, T., Doya, K., Toyama, K., Kimura, M., Samejima, K., Imamizu, H., & Kawato, M. (2004). A neural correlate of reward-based behavioral learning in caudate nucleus: A functional magnetic resonance imaging study of a stochastic decision task. *Journal of Neuroscience, 24,* 1660–1665.

Hauser, M. D., Chomsky, N., & Fitch, W. T. (2002). The faculty of language: What is it, who has it, and how did it evolve? *Science, 298,* 1569–1579.

Hayward, W. G., & Tarr, M. J. (1997). Testing conditions for viewpoint invariance in object recognition. *Journal of Experimental Psychology: Human Perception and Performance, 23,* 1511–1521.

Hebb, D. O. (2002). *The Organization of Behavior: A Neuropsychological Theory.* Mahwah, NJ: Lawrence Erlbaum. (Original work published 1949.)

Heekeren, H. R., Marrett, S., Ruff, D. A., Bandettini, P. A., & Ungerleider, L. G. (2006). Involvement of human left dorsolateral prefrontal cortex in perceptual decision making is independent of response modality. *Proceedings of the National Academy of Sciences of the United States of America, 103,* 10023–10028.

Heidegger, M. (2003). *The End of Philosophy.* (J. Stambaugh, Transl.) Chicago, IL: University of Chicago Press. (Original work published 1954, 1961; translation first published 1973.)

Hejinian, L. (2000). *The Language of Inquiry*. Berkeley, CA: University of California Press.

Henderson, J. M., Weeks, P. A., & Hollingworth, A. (1999). The effects of semantic consistency on eye movements during complex scene viewing. *Journal of Experimental Psychology: Human Perception and Performance, 25,* 210–228.

Herrnstein, R. J. (1961). Relative and absolute strength of responses as a function of frequency of reinforcement. *Journal of the Experimental Analysis of Behavior, 4,* 267–272.

Herrnstein, R. J., & Murray, C. (1996). *The Bell Curve: Intelligence and Class Structure in American Life*. New York, NY: Free Press. (Original work published 1994.)

Herry, C., Ciocchi, S., Senn, V., Demmou, L., Müller, C., & Lüthi, A. (2008). Switching on and off fear by distinct neuronal circuits. *Nature, 454,* 600–606.

Hikosaka, O. (1999). Probing consciousness with an electrode. *Brain Research Bulletin, 50,* 337–338.

Hikosaka, O., & Wurtz, R. H. (1983). Visual and oculomotor functions of monkey substantia nigra pars reticulata. III. Memory-contingent visual and saccade responses. *Journal of Neurophysiology, 1983,* 1268–1284.

Hikosaka, O., Sakamoto, M., & Usui, S. (1989). Functional properties of monkey caudate neurons. III. Activities related to expectation of target and reward. *Journal of Neurophysiology, 61,* 814–832.

Hikosaka, O., Takikawa, Y., & Kawagoe, R. (2000). Role of the basal ganglia in the control of purposive saccadic eye movements. *Physiological Reviews, 80,* 953–978.

Hofstadter, D. R. (1999). *Gödel, Escher, Bach: An Eternal Golden Braid*. New York, NY: Basic Books. (Original work published 1979.)

Hollerman, J. R., & Schultz, W. (1998). Dopamine neurons report an error in the temporal prediction of reward during learning. *Nature Neuroscience, 1,* 304–309.

Hollingworth, A., & Henderson, J. M. (1998). Does consistent scene context facilitate object perception? *Journal of Experimental Psychology: General, 127,* 398–415.

Hong, S., & Hikosaka, O. (2008). The globus pallidus sends reward-related signals to the lateral habenula. *Neuron, 60,* 720–729.

Houweling, A. R., & Brecht, M. (2008). Behavioural report of single neuron stimulation in somatosensory cortex. *Nature, 451,* 65–68.

Howe, S. (2007). *Souls of the Labadie Tract*. New York, NY: New Directions Books.

Hubel, D. H., & Wiesel, T. N. (1959). Receptive fields of single neurones in the cat's striate cortex. *Journal of Physiology, 148,* 574–591.

Huettel, C. A., Stowe, C. J., Gordon, E. M., Warner, B. T., & Platt, M. L. (2006). Neural signatures of economic preferences for risk and ambiguity. *Neuron, 49,* 765–775.

Hume, D. (2002). *A Treatise of Human Nature*. (D. F. Norton & M. J. Norton, Eds.) Oxford, U.K.: Oxford University Press. (Original work published 1740.)

Huxley, A. (2004). *The Doors of Perception & Heaven and Hell*. New York, NY: HarperCollins. (*The Doors of Perception* first published 1954; *Heaven and Hell* first published 1956.)

Hsu, M., Anen, C., & Quartz, S. R. (2008). The right and the good: Distributive justice and neural encoding of equity and efficiency. *Science, 320,* 1092–1095.

Hwang, W. S., Roh, S. I., Lee, B. C., Kang, S. K., Kwon, D. K., Kim, S., Kim, S. J., Park, S. W., Kwon, H. S., Lee, C. K., Lee, J. B., Kim, J. M., Ahn, C., Paek, S. H., Chang, S. S., Koo, J. J., Yoon, H. S., Hwang, J. H., Hwang, Y. Y., Park, Y. S., Oh, S. K., Kim, H. S., Park, J. H., Moon, S. Y., & Schatten, G. (2005). Patient-specific embryonic stem cells derived from human SCNT blastocysts. *Science, 308,* 1777–1783.

Hwang, W. S., Ryu, Y. J., Park, J. H., Park, E. S., Lee, E. G., Koo, J. M., Jeon, H. Y., Lee, B. C., Kang, S. K., Kim, S. J., Ahn, C., Hwang, J. H., Park, K. Y., Cibelli, J. B., & Moon, S. Y. (2004). Evidence of a pluripotent human embryonic stem cell line derived from a cloned blastocyst. *Science, 303,* 1669–1674.

Ikeda, T., & Hikosaka, O. (2003). Reward-dependent gain and bias of visual responses in primate superior colliculus. *Neuron, 39,* 693–700.

Isbell, L. A. (2006). Snakes as agents of evolutionary change in primate brains. *Journal of Human Evolution, 51,* 1–35.

Ito, J., Nikolaev, A. R., & Van Leeuwen, C. (2007). Dynamics of spontaneous transitions between global brain states. *Human Brain Mapping, 28,* 904–913.

Izuma, K., Saito, D. N., & Sadato, N. (2008). Processing of social and monetary rewards in the human striatum. *Neuron, 58,* 284–294.

Jackendoff, R. (2007). *Language, Consciousness, Culture: Essays on Mental Structure.* Cambridge, MA: MIT Press.

James, W. (1950). *The Principles of Psychology.* Volume One. New York, NY: Dover Publications. (Original work published 1890.)

Johansson, P., Hall, L., Sikström, S., & Olsson, A. (2005). Failure to detect mismatches between intention and outcome in a simple decision task. *Science, 310,* 116–119.

Jonides, J. (1981). Voluntary versus automatic control over the mind's eye's movement. In: J. B. Long & A. D. Baddeley (Eds.), *Attention and Performance IX* (pp. 187–203). Hillsdale, NJ: Lawrence Erlbaum.

Kahneman, D., & Tversky, A. (1972). Subjective probability: A judgment of representativeness. *Cognitive Psychology, 3,* 430–454.

Kahneman, D., & Tversky, A. (1979). Prospect theory: An analysis of decision under risk. *Econometrica, 47,* 263–291.

Kahneman, D., & Tversky, A. (Eds.) (2000). *Choices, Values, and Frames.* Cambridge, U.K.: Cambridge University Press.

Kahneman, D., Slovic, P., & Tversky, A. (Eds.) (1982). *Judgement under Uncertainty: Heuristics and Biases.* Cambridge, U.K.: Cambridge University Press.

Kastner, S., Pinsk, M. A., De Weerd, P., Desimone, R., & Ungerleider, L. G. (1999). Increased activity in human visual cortex during directed attention in the absence of visual stimulation. *Neuron, 22,* 751–761.

Kawagoe, R., Takikawa, Y., & Hikosaka, O. (1998). Expectation of reward modulates cognitive signals in the basal ganglia. *Nature Neuroscience, 1,* 411–416.

Kawagoe, R., Takikawa, Y., & Hikosaka, O. (2004). Reward-predicting activity of dopamine and caudate neurons—A possible mechanism of motivational control of saccadic eye movement. *Journal of Neurophysiology, 91,* 1013–1024.

Keats, J. (1974). *Poems.* (G. Bullet, Ed.) London, U.K.: J. M. Dent & Sons (Everyman's Library). (Edition first published 1906.)

Kelemen, D. (2004). Are children "intuitive theists"? Reasoning about purpose and design in nature. *Psychological Science, 15,* 295–301.

Kendall, M. G., & Buckland, W. R. (1957). *A Dictionary of Statistical Terms.* Edinburgh, U.K.: Tweeddale Court.

Kenkō, Y. (2001). *Essays in Idleness: The Tsurezuregusa of Kenkō.* (D. Keene, Transl.) Tokyo, Japan: Tuttle Publishing. (Translation first published 1967.)

Kennedy, D. (2006). Editorial retraction. *Science, 311,* 335.

Kepecs, A., Uchida, S., Zariwala, H. A., & Mainen, Z. F. (2008). Neural correlates, computation and behavioural impact of decision confidence. *Nature, 455,* 227–231.

Kim, H., & Cabeza, R. (2007). Trusting our memories: Dissociating the neural correlates of confidence in veridical versus illusory memories. *Journal of Neuroscience, 27,* 12190–12197.

Kim, J. N., & Shadlen, M. N. (1999). Neural correlates of a decision in the dorsolateral prefrontal cortex of the macaque. *Nature Neuroscience, 2,* 176–185.

Kim, Y. J., Grabowecky, M., Paller, K. A., Muthu, K., & Suzuki, S. (2007). Attention induces synchronization-based response gain in steady-state visual evoked potentials. *Nature Neuroscience, 10,* 117–125.

King-Casas, B., Tomlin, D., Anen, C., Camerer, C. F., Quartz, S. R., & Montague, P. R. (2005). Getting to know you: Reputation and trust in a two-person economic exchange. *Science, 308,* 78–83.

Knetsch, J. L., & Sinden, J. A. (1984). Willingness to pay and compensation demanded: Experimental evidence of an unexpected disparity in measures of value. *Quarterly Journal of Economics, 99,* 507–521.

Knutson, K. M., Mah, L., Manly, C. F., & Grafman, J. (2007). Neural correlates of automatic beliefs about gender and race. *Human Brain Mapping, 28,* 915–930.

Kobayashi, S., & Schultz, W. (2008). Influence of reward delays on responses of dopamine neurons. *Journal of Neuroscience, 28,* 7837–7846.

Kobayashi, S., Kawagoe, R., Takikawa, Y., Koizumi, M., Sakagami, M., & Hikosaka, O. (2007). Functional differences between macaque prefrontal cortex and caudate nucleus during eye movements with and without reward. *Experimental Brain Research, 176,* 341–355.

Kobayashi, S., Lauwereyns, J., Koizumi, M., Sakagami, M., & Hikosaka, O. (2002). Influence of reward expectation on visuospatial processing in macaque lateral prefrontal cortex. *Journal of Neurophysiology, 87,* 1488–1498.

Kobayashi, S., Nomoto, K., Watanabe, M., Hikosaka, O., Schultz, W., & Sakagami, M. (2006). Influences of rewarding and aversive outcomes on activity in macaque lateral prefrontal cortex. *Neuron, 51,* 861–870.

Koivisto, M., & Revonsuo, A. (2007). How meaning shapes seeing. *Psychological Science, 18,* 845–849.

Kornblum, S., Hasbroucq, T., & Osman, A. (1990). Dimensional overlap: Cognitive basis for stimulus–response compatibility—A model and a taxonomy. *Psychological Review, 97,* 253–270.

Krauzlis, R. J., & Dill, N. (2002). Neural correlates of target choice for pursuit and saccades in the primate superior colliculus. *Neuron, 35,* 355–363.

Krawczyk, D. C., Gazzaley, A., & D'Esposito, M. (2007). Reward modulation of prefrontal and visual association cortex during an incentive working memory task. *Brain Research, 1141,* 168–177.

Krekelberg, B., Boynton, G. M., & Van Wezel, R. J. (2006). Adaptation: From single cells to BOLD signals. *Trends in Neurosciences, 29,* 250–256.

Krueger, L. E., & Shapiro, R. G. (1981). A reformulation of Proctor's unified theory for matching-task phenomena. *Psychological Review, 88,* 573–581.

LaBerge, D. (1983). Spatial extent of attention to letters and words. *Journal of Experimental Psychology: Human Perception and Performance, 9,* 371–379.

Lacan, J. (2004). *Écrits: A Selection.* (A. Sheridan, Transl.) London, U.K.: Routledge. (Original work published 1966; translation first published 1977.)

Lammertyn, J., Fias, W., & Lauwereyns, J. (2002). Semantic influences on feature-based attention due to overlap of neural circuits. *Cortex, 38,* 878–882.

LaMotte, R. H., & Mountcastle, V. B. (1975). Capacities of humans and monkeys to discriminate between the vibratory stimuli of different frequency and amplitude: A correlation between neural events and psychophysical measurements. *Journal of Neurophysiology, 38,* 539–559.

Lashley, K. S. (1931). Mass action in cerebral function. *Science, 73,* 245–254.

Lauwereyns, J. (1998). Exogenous/endogenous control of space-based/object-based attention: Four types of visual selection? *European Journal of Cognitive Psychology, 10,* 41–74.

Lauwereyns, J. (2006). Voluntary control of unavoidable action. *Trends in Cognitive Sciences, 10,* 47–49.

Lauwereyns, J. (2008). The contribution of dopamine to the implementation of reward value during the control of action. *Central Nervous System Agents in Medicinal Chemistry, 8,* 72–84.

Lauwereyns, J., & d'Ydewalle, G. (1996). Knowledge acquisition in poetry criticism: The expert's eye movements as an information tool. *International Journal of Human–Computer Studies, 45,* 1–18.

Lauwereyns, J., & d'Ydewalle, G. (1997). Global orientation disrupts the detection of a similar local orientation. *Perception, 26,* 1259–1270.

Lauwereyns, J., & Wisnewski, R. G. (2006). A reaction-time paradigm to measure reward-oriented bias in rats. *Journal of Experimental Psychology: Animal Behavior Processes, 32,* 467–473.

Lauwereyns, J., Koizumi, M., Sakagami, M., Hikosaka, O., Kobayashi, S., & Tsutsui, K. (2000). Interference from irrelevant features on visual discrimination by macaques (*Macaca fuscata*): A behavioral analogue of the human Stroop effect. *Journal of Experimental Psychology: Animal Behavior Processes, 26,* 352–357.

Lauwereyns, J., Sakagami, M., Tsutsui, K., Kobayashi, S., Koizumi, M., & Hikosaka, O. (2001). Responses to task-irrelevant visual features by primate prefrontal neurons. *Journal of Neurophysiology, 86,* 2001–2010.

Lauwereyns, J., Takikawa, Y., Kawagoe, R., Kobayashi, S., Koizumi, M., Coe, B., Sakagami, M., & Hikosaka, O. (2002a). Feature-based anticipation of cues that predict reward in monkey caudate nucleus. *Neuron, 33,* 463–473.

Lauwereyns, J., Watanabe, K., Coe, B., & Hikosaka, O. (2002b). A neural correlate of response bias in monkey caudate nucleus. *Nature, 418,* 413–417.

Lauwereyns, J., Wisnewski, R., Keown, K., & Govan, S. (2006). Crosstalk between on-line and off-line processing of visual features. *Psychological Research, 70,* 170–179.

Lavie, N., & Tsal, Y. (1994). Perceptual load as a major determinant of the locus of selection in visual attention. *Perception & Psychophysics, 56,* 183–197.

LeDoux, J. E. (1994). Emotion, memory and the brain. *Scientific American, 270,* 50–57.

LeDoux, J. E. (2002). *Synaptic Self: How Our Brains Become Who We Are.* New York, NY: Viking.

Lee, B. C., Kim, M. K., Jang, G., Oh, H. J., Yuda, F., Kim, H. J., Hossein, M. S., Kim, J. J., Kang, S. K., Schatten, G., & Hwang, W. S. (2005). Dogs cloned from adult somatic cells. *Nature, 436,* 641.

Leon, M. I., & Shadlen, M. N. (1999). Effect of expected reward magnitude on the response of neurons in the dorsolateral prefrontal cortex of the macaque. *Neuron, 24,* 415–425.

Leon, M. I., Poytress, B. S., & Weinberger, N. M. (2008). Avoidance learning facilitates temporal processing in the primary auditory cortex. *Neurobiology of Learning and Memory, 90,* 347–357.

Lewis, P. A., Critchley, H. D., Rotshtein, P., & Dolan, R. J. (2007). Neural correlates of processing valence and arousal in affective words. *Cerebral Cortex, 17,* 724–748.

Li, W., Howard, J. D., Parrish, T. B., & Gottfried, J. A. (2008). Aversive learning enhances perceptual and cortical discrimination of indiscriminable odor cues. *Science, 319,* 1842–1845.

Lindsay, D. S., Hagen, L., Read, J. D., Wade, K. A., & Garry, M. (2004). True photographs and false memories. *Psychological Science, 15,* 149–154.

Liston, D. B., & Stone, L. S. (2008). Effects of prior information and reward on oculomotor and perceptual choices. *Journal of Neuroscience, 28,* 13866–13875.

Liu, T., Larsson, J., & Carrasco, M. (2007). Feature-based attention modulates orientation-selective responses in human visual cortex. *Neuron, 55,* 313–323.

Liu, Z., Richmond, R. J., Murray, E. A., Saunders, R. C., Steenrod, S., Stubblefield, B. K., Montague, D. M., & Ginns, E. I. (2004). DNA targeting of rhinal cortex D2 receptor protein reversibly blocks learning of cues that predict reward. *Proceedings of the National Academy of Sciences of the United States of America, 101,* 12336–12341.

Lo, C. C., & Wang, X. J. (2006). Cortico–basal ganglia circuit mechanism for a decision threshold in reaction time tasks. *Nature Neuroscience, 9,* 956–963.

Loftus, E. F. (2003). Make-believe memories. *American Psychologist, 58,* 867–873.

Logan, G. D. (1996). The CODE theory of visual attention: An integration of space-based and object-based attention. *Psychological Review, 103,* 603–649.

Logothetis, N. K. (2008). What we can do and what we cannot do with fMRI. *Nature, 453,* 869–878.

Lombrozo, T., Kelemen, D., & Zaitchik, D. (2007). Inferring design: Evidence of a preference for teleological explanations in patients with Alzheimer's disease. *Psychological Science, 18,* 999–1006.

Lu, X., Matsuzawa, M., & Hikosaka, O. (2002). A neural correlate of oculomotor sequences in supplementary eye field. *Neuron, 34,* 317–325.

Lucas, C., & Lauwereyns, J. (2007). Selective working memory disables inhibition of visual features. *Experimental Psychology, 54,* 256–263.

Luce, R. D. (1963). A threshold theory for simple detection experiments. *Psychological Review, 70,* 61–79.

Luce, R. D. (1986). *Response Times: Their Role in Inferring Elementary Mental Organization.* Oxford, U.K.: Clarendon Press.

Luce, R. D., & Green, D. M. (1972). A neural timing theory for response times and the psychophysics of intensity. *Psychological Review, 79,* 14–57.

Luck, S. J., Chelazzi, L., Hillyard, S. A., & Desimone, R. (1997). Neural mechanisms of spatial selective attention in areas V1, V2, and V4 of macaque visual cortex. *Journal of Neurophysiology, 77,* 24–42.

MacArthur, B. (Ed.) (1999). *The Penguin Book of Twentieth-Century Speeches.* London, U.K.: Penguin Books. (Second revised edition.)

MacDonald, A. W. 3rd, Cohen, J. D., Stenger, V. A., & Carter, C. S. (2000). Dissociating the role of the dorsolateral prefrontal and anterior cingulate cortex in cognitive control. *Science, 288,* 1835–1838.

MacIntyre, A. (2004). *A Short History of Ethics.* London, U.K.: Routledge. (Original work published 1967.)

Mack, A., & Rock, I. (1998). *Inattentional Blindness.* Cambridge, MA: MIT Press.

Mackworth, N. H., & Morandi, A. J. (1967). The gaze selects informative details within pictures. *Perception & Psychophysics, 2,* 547–552.

MacLeod, C. M. (1991a). Half a century of research on the Stroop effect: An integrative review. *Psychological Bulletin, 109,* 163–203.

MacLeod, C. M. (1991b). John Ridley Stroop: Creator of a landmark cognitive task. *Canadian Psychology, 32,* 521–524.

MacLeod, C. M. (1992). The Stroop task: The "gold standard" of attentional measures. *Journal of Experimental Psychology: General, 121,* 12–14.

Majewska, A. K., Newton, J. R., & Sur, M. (2006). Remodeling of synaptic structure in sensory cortical areas *in vivo. Journal of Neuroscience, 26,* 3021–3029.

Mamassian, P. (2006). Bayesian inference of form and shape. *Progress in Brain Research, 154,* 265–270.

Mamassian, P., & Landy, M. S. (1998). Observer biases in the 3D interpretation of line drawings. *Vision Research, 38,* 2817–2832.

Manyōshū. (2005). *1000 Poems from the Manyōshū.* (Japanese Classics Translation Committee, Transl.) Mineola, NY: Dover Publications. (Translation first published 1940.)

Marr, D. (1982). *Vision: A Computational Investigation into the Human Representation and Processing of Visual Information.* New York, NY: W.H. Freeman and Company.

Matsumoto, M., & Hikosaka, O. (2007). Lateral habenula as a source of negative reward signals in dopamine neurons. *Nature, 447,* 1111–1115.

Mattingley, J. B., Davis, G., & Driver, J. (1997). Preattentive filling-in of visual surfaces in parietal extinction. *Science, 275,* 671–674.

Maunsell, J. H. R. (2004). Neuronal representations of cognitive state: Reward or attention? *Trends in Cognitive Sciences, 8,* 261–265.

McAdams, C. J., & Maunsell, J. H. R. (1999). Effects of attention on orientation-tuning functions of single neurons in macaque cortical area V4. *Journal of Neuroscience, 19,* 431–441.

McCabe, K., Houser, D., Ryan, L., Smith, V., & Trouard, T. (2001). A functional imaging study of cooperation in two-person reciprocal exchange. *Proceedings of the National Academy of Sciences of the United States of America, 98,* 11832–11835.

McMahon, D. B. T., & Olson, C. R. (2007). Repetition suppression in monkey inferotemporal cortex: Relation to behavioral priming. *Journal of Neurophysiology, 97,* 3532–3543.

Melcher, D. (2007). Predictive remapping of visual features precedes saccadic eye movements. *Nature Neuroscience, 10,* 903–907.

Melloni, L., Molina, C., Pena, M., Torres, D., Singer, W., & Rodriguez, E. (2007). Synchronization of neural activity across cortical areas correlates with conscious perception. *Journal of Neuroscience, 27,* 2858–2865.

Miller, E. K. (1999). Neurobiology: Straight from the top. *Nature, 401,* 650–651.

Miller, G. A. (1956). The magical number seven plus or minus two: Some limits on our capacity for processing information. *Psychological Review, 63,* 81–97.

Milton, J. (1989). *Paradise Lost.* (C. Ricks, Ed.) London, U.K.: Penguin Books. (Original work published 1667.)

Minamimoto, T., & Kimura, M. (2002). Participation of the thalamic CM-Pf complex in attentional orienting. *Journal of Neurophysiology, 87,* 3090–3101.

Minamimoto, T., Hori, Y., & Kimura, M. (2005). Complementary process to response bias in the centromedian nucleus of the thalamus. *Science, 308,* 1798–1801.

Miyashita, K., Rand, M. K., Miyachi, S., & Hikosaka, O. (1996). Anticipatory saccades in sequential procedural learning in monkeys. *Journal of Neurophysiology, 76,* 1361–1366.

Mobbs, D., Petrovic, P., Marchant, J. L., Hassabis, D., Weiskopf, N., Seymour, B., Dolan, R. J., & Frith, C. D. (2007). When fear is near: Threat imminence elicits prefrontal-periaqueductal gray shifts in humans. *Science, 317,* 1079–1083.

Mogenson, G. J., Jones, D. L., & Yim, C. Y. (1980). From motivation to action: Functional interface between the limbic system and the motor system. *Progress in Neurobiology, 14,* 69–97.

Mongillo, G., Barak, O., & Tsodyks, M. (2008). Synaptic theory of working memory. *Science, 319,* 1543–1546.

Montefiore, S. S. (Ed.) (2008). *Speeches That Changed the World: The Stories and Transcripts of the Moments That Made History.* Essex, U.K.: Quercus.

More, T. (2003). *Utopia.* (P. Turner, Transl.) London, U.K.: Penguin. (Original work published 1516; translation first published 1965.)

Morris, J. S., Buchel, C., & Dolan, R. J. (2001). Parallel neural responses in amygdala subregions and sensory cortex during implicit fear conditioning. *Neuroimage, 13,* 1044–1052.

Mountcastle, V. B., LaMotte, R. H., & Carli, G. (1972). Detection thresholds for stimuli in humans and monkeys: Comparison with threshold events in mechanoreceptive afferent nerve fibers innervating the monkey hand. *Journal of Neurophysiology, 35,* 122–136.

Mruczek, R. E. B., & Sheinberg, D. L. (2007). Context familiarity enhances target processing by inferior temporal cortex neurons. *Journal of Neuroscience, 27,* 8533–8545.

Munoz, D. P., & Everling, S. (2004). Look away: The anti-saccade task and the voluntary control of eye movement. *Nature Reviews Neuroscience, 5,* 218–228.

Munoz, D. P., & Wurtz, R. H. (1995). Saccade-related activity in monkey superior colliculus. I. Characteristics of burst and buildup cells. *Journal of Neurophysiology, 73,* 2313–2333.

Nagano-Saito, A., Leyton, M., Monchi, O., Goldberg, Y. K., He, Y., & Dagher, A. (2008). Dopamine depletion impairs frontostriatal functional connectivity during a set-shifting task. *Journal of Neuroscience, 28,* 3697–3706.

Nakahara, H., Itoh, H., Kawagoe, R., Takikawa, Y., & Hikosaka, O. (2004). Dopamine neurons can represent context-dependent prediction error. *Neuron, 41,* 269–280.

Nakahara, H., Nakamura, K., & Hikosaka, O. (2006). Extended LATER model can account for trial-by-trial variability of both pre- and post-processes. *Neural Networks, 19,* 1027–1046.

Nakamura, K., & Hikosaka, O. (2006a). Facilitation of saccadic eye movements by postsaccadic electrical stimulation in the primate caudate. *Journal of Neuroscience, 26,* 12885–12895.

Nakamura, K., & Hikosaka, O. (2006b). Role of dopamine in the primate caudate nucleus in reward modulation of saccades. *Journal of Neuroscience, 26,* 5360–5369.

Nakamura, K., Roesch, M. R., & Olson, C. R. (2005). Neuronal activity in macaque SEF and ACC during performance of tasks involving conflict. *Journal of Neurophysiology, 93,* 884–908.

Newsome, W. T., Britten, K. H., & Movshon, J. A. (1989). Neuronal correlates of a perceptual decision. *Nature, 341,* 52–54.

Nicola, S. M., Surmeier, J., & Malenka, R. C. (2000). Dopaminergic modulation of neuronal excitability in the striatum and nucleus accumbens. *Annual Review of Neuroscience, 23,* 185–215.

Nieoullon, A. (2002). Dopamine and the regulation of cognition and attention. *Progress in Neurobiology, 67,* 53–83.

O'Doherty, J., Kringelbach, M. L., Rolls, E. T., Hornak, J., & Andrews, C. (2001). Abstract reward and punishment representations in the human orbitofrontal cortex. *Nature Neuroscience, 4,* 95–102.

Öhman, A. (2007). Has evolution primed humans to "beware the beast"? *Proceedings of the National Academy of Sciences of the United States of America, 104,* 16396–16397.

Öhman, A., & Mineka, S. (2001). Fears, phobias, and preparedness: Toward an evolved module of fear and fear learning. *Psychological Review, 108,* 483–522.

Öhman, A., Flykt, A., & Esteves, F. (2001). Emotion drives attention: Detecting the snake in the grass. *Journal of Experimental Psychology: General, 130,* 466–478.

Olds, J., & Milner, P. (1954). Positive reinforcement produced by electrical stimulation of septal area and other regions of rat brain. *Journal of Comparative and Physiological Psychology, 47,* 419–427.

Op de Beeck, H. P., Torfs, K., & Wagemans, J. (2008). Perceived shape similarity among unfamiliar objects and the organization of the human object vision pathway. *Journal of Neuroscience, 28,* 10111–10123.

Oxley, D. R., Smith, K. B., Alford, J. R., Hibbing, M. V., Miller, J. L., Scalora, M., Hatemi, P. K., & Hibbing, J. R. (2008). Political attitudes vary with physiological traits. *Science, 321,* 1667–1670.

Palmer, M. (1998). *The Lion Bridge: Selected Poems 1972–1995.* New York, NY: New Directions Books.

Palmer, M. (2008). *Active Boundaries: Selected Essays and Talks.* New York, NY: New Directions Books.

Pan, W., Schmidt, R., Wickens, J. R., & Hyland, B. I. (2005). Dopamine cells respond to predicted events during classical conditioning: Evidence for eligibility traces in the reward-learning network. *Journal of Neuroscience, 25,* 6235–6242.

Pan, X., Sawa, K., Tsuda, I., Tsukada, M., & Sakagami, M. (2008). Reward prediction based on stimulus categorization in primate lateral prefrontal cortex. *Nature Neuroscience, 11,* 703–712.

Paré, M., & Hanes, D. P. (2003). Controlled movement processing: Superior colliculus activity associated with countermanded saccades. *Journal of Neuroscience, 23,* 6480–6489.

Pashler, H. (1984). Processing stages in overlapping tasks: Evidence for a central bottleneck. *Journal of Experimental Psychology: Human Perception and Performance, 10,* 358–377.

Pashler, H. (1994). Graded capacity sharing in dual-task interference? *Journal of Experimental Psychology: Human Perception and Performance, 20,* 330–342.

Pasupathy, A., & Miller, E. K. (2005). Different time courses of learning-related activity in the prefrontal cortex and striatum. *Nature, 433,* 873–876.

Penrose, R. (1989). *The Emperor's New Mind.* Oxford, U.K.: Oxford University Press.

Pessiglione, M., Schmidt, L., Draganski, B., Kalisch, R., Lau, H., Dolan, R. J., & Frith, C. D. (2007). How the brain translates money into force: A neuroimaging study of subliminal motivation. *Science, 316,* 904–906.

Pessoa, F. (2002). *The Book of Disquiet.* (R. Zenith, Transl.) London, U.K.: Penguin Classics. (Translation first published 2000; Original work published 1998.)

Phelps, E. A., & LeDoux, J. E. (2005). Contributions of the amygdala to emotion processing: From animal models to human behavior. *Neuron, 48,* 175–187.

Pinel, P., Piazza, M., Le Bihan, D., & Dehaene, S. (2004). Distributed and overlapping cerebral representations of number, size, and luminance during comparative judgments. *Neuron, 41,* 983–993.

Pinker, S. (1999). *How the Mind Works.* New York, NY: W.W. Norton. (Original work published 1997.)

Pinker, S. (2000). *The Language Instinct: How the Mind Creates Language.* New York, NY: HarperCollins. (Original work published 1994.)

Pinker, S. (2002). *The Blank Slate: The Modern Denial of Human Nature.* New York, NY: Viking Penguin.

Platt, M. L., & Glimcher, P. W. (1999). Neural correlates of decision variables in parietal cortex. *Nature, 400,* 233–238.

Pleger, B., Blankenburg, F., Ruff, C. C., Driver, J., & Dolan, R. J. (2008). Reward facilitates tactile judgments and modulates hemodynamic responses in human primary somatosensory cortex. *Journal of Neuroscience, 28,* 8161–8168.

Plous, S. (1993). *The Psychology of Judgment and Decision Making.* New York, NY: McGraw-Hill.

Pochon, J. B., Riis, J., Sanfey, A. G., Nystrom, L. E., & Cohen, J. D. (2008). Functional imaging of decision conflict. *Journal of Neuroscience, 28,* 3468–3473.

Polk, T. A., Drake, R. M., Jonides, J. J., Smith, M. R., & Smith, E. E. (2008). Attention enhances the neural processing of relevant features and suppresses the processing of irrelevant features in humans: A functional magnetic resonance imaging study of the Stroop task. *Journal of Neuroscience, 28,* 13786–13792.

Popper, K. (2002a). *The Logic of Scientific Discovery.* London, U.K.: Routledge. (Original work published 1935; translation by the author first published 1959.)

Popper, K. (2002b). *The Open Society and Its Enemies.* London, U.K.: Routledge. (Original work published 1945; translation by the author first published 1962.)

Posner, M. I. (1980). Orienting of attention. *Quarterly Journal of Experimental Psychology, 32,* 3–25.

Pringle, H. L., Irwin, D. E., Kramer, A. F., & Atchley, P. (2001). The role of attentional breadth in perceptual change detection. *Psychonomic Bulletin & Review, 8,* 89–95.

Proctor, R. W. (1981). A unified theory for matching-task phenomena. *Psychological Review, 88,* 291–326.

Proctor, R. W., & Rao, K. V. (1983). Reinstating the original principles of Proctor's unified theory for matching-task phenomena: An evaluation of Krueger and Shapiro's reformulation. *Psychological Review, 90,* 21–37.

Prynne, J. H. (2005). *Poems.* Northumberland, U.K.: Bloodaxe Books.

Quattrone, G. A., & Tversky, A. (1988). Contrasting rational and psychological analyses of political choice. *American Political Science Review, 82,* 719–736.

Quine, W. V. (1966). *The Ways of the Paradox and Other Essays.* New York, NY: Random House.

Quiroga, R. Q., Reddy, L., Kreiman, G., Koch, C., & Fried, I. (2005). Invariant visual representation by single neurons in the human brain. *Nature, 435,* 1036–1037.

Rainer, G., Rao, S. C., & Miller, E. K. (1999). Prospective coding for objects in primate prefrontal cortex. *Journal of Neuroscience, 19,* 5493–5505.

Rao, R. P. N., Olshausen, B. A., & Lewicki, M. S. (Eds.) (2002). *Probabilistic Models of the Brain: Perception and Neural Function.* Cambridge, MA: MIT Press.

Ratcliff, R., Hasegawa, Y. T., Hasegawa, R. P., Smith, P. L., & Segraves, M. A. (2007). Dual diffusion model for single-cell recording data from the superior colliculus in a brightness-discrimination task. *Journal of Neurophysiology, 97,* 1756–1774.

Ratcliff, R., Van Zandt, T., & McKoon, G. (1999). Connectionist and diffusion models of reaction time. *Psychological Review, 106,* 261–300.

Reddi, B. A. J., & Carpenter, R. H. S. (2000). The influence of urgency on decision time. *Nature Neuroscience, 3,* 827–830.

Reddi, B. A. J., Assress, K. N., & Carpenter, R. H. S. (2003). Accuracy, information, and response time in a saccadic decision task. *Journal of Neurophysiology, 90,* 3538–3546.

Redgrave, P., & Gurney, K. (2006). The short-latency dopamine signal: A role in discovering novel actions? *Nature Reviews Neuroscience, 7,* 967–975.

Redgrave, P., Prescott, T. J., & Gurney, K. (1999). Is the short-latency dopamine response too short to signal reward error? *Trends in Neurosciences, 22,* 146–151.

Redish, A. D. (2004). Addiction as a computational process gone awry. *Science, 306,* 1944–1947.

Rescorla, R. A., & Wagner, A. R. (1972). A theory of Pavlovian conditioning: Variations in the effectiveness of reinforcement and nonreinforcement. In: A. H. Black & W. F. Prokasy (Eds.), *Classical Conditioning II: Current Research and Theory* (pp. 64–99). New York, NY: Appleton Century Crofts.

Reynolds, J. N. J., Hyland, B. I., & Wickens, J. R. (2001). A cellular mechanism of reward-related learning. *Nature, 413,* 67–70.

Rhodes, G. (2006). The evolutionary psychology of facial beauty. *Annual Review of Psychology, 57,* 199–226.

Ridley, M. (2003). *Nature via Nurture: Genes, Experience, and What Makes Us Human.* New York, NY: HarperCollins.

Rizzolatti, G., Riggio, L., Dascola, I., & Umiltá, C. (1987). Reorienting attention across the horizontal and vertical meridians: Evidence in favor of a premotor theory of attention. *Neuropsychologia, 25,* 31–40.

Roberts, A. D., Robbins, T. W., & Weiskrantz, L. (Eds.) (1998). *The Prefrontal Cortex: Executive and Cognitive Functions.* Oxford, U.K.: Oxford University Press.

Rodriguez, E., George, N., Lachaux, J. P., Martinerie, J., Renault, B., & Varela, F. J. (1999). Perception's shadow: Long-distance synchronization of human brain activity. *Nature, 397,* 430–433.

Roelfsema, P. R. (2006). Cortical algorithms for perceptual grouping. *Annual Review of Neuroscience, 29,* 203–227.

Roesch, M. R., & Olson, C. R. (2003). Impact of expected reward on neuronal activity in prefrontal cortex, frontal and supplementary eye fields and premotor cortex. *Journal of Neurophysiology, 90,* 1766–1789.

Roesch, M. R., & Olson, C. R. (2004). Neuronal activity related to reward value and motivation in primate frontal cortex. *Science, 304,* 307–310.

Roesch, M. R., Calu, D. J., & Schoenbaum, G. (2007). Dopamine neurons encode the better option in rats deciding between differently delayed or sized rewards. *Nature Neuroscience, 10,* 1615–1624.

Roiser, J. P., Levy, J., Fromm, S. J., Wang, H., Hasler, G., Sahakian, B. J., & Drevets, W. C. (2008). The effect of acute tryptophan depletion on the neural correlates of emotional processing in healthy volunteers. *Neuropsychopharmacology, 33,* 1992–2006.

Roitman, J. D., & Shadlen, M. N. (2002). Response of neurons in the lateral intraparietal area during a combined visual discrimination task. *Journal of Neuroscience, 22,* 9475–9489.

Rolls, E. T. (2008). *Memory, Attention, and Decision-Making: A Unifying Computational Neuroscience Approach.* Oxford, U.K.: Oxford University Press.

Rossi, A. F., Bichot, N. P., Desimone, R., & Ungerleider, L. G. (2007). Top down attentional deficits in macaques with lesions of lateral prefrontal cortex. *Journal of Neuroscience, 27,* 11306–11314.

Royce, G. J., Bromley, S., & Gracco, C. (1991). Subcortical projections to the centromedian and parafascicular nuclei in the cat. *Journal of Comparative Neurology, 306,* 129–155.

Russell, J. A. (1991). Culture and the categorization of emotions. *Psychological Bulletin, 110,* 426–450.

Saalmann, Y. B., Pigarev, I. N., & Vidyasagar, T. R. (2007). Neural mechanisms of visual attention: How top–down feedback highlights relevant locations. *Science, 316,* 1612–1615.

Sakagami, M., & Niki, H. (1994). Encoding of behavioral significance of visual stimuli by primate prefrontal neurons: Relation to relevant task conditions. *Experimental Brain Research, 97,* 423–436.

Sakagami, M., & Tsutsui, K. (1999). The hierarchical organization of decision making in the primate prefrontal cortex. *Neuroscience Research, 34,* 79–89.

Sakagami, M., Tsutsui, K., Lauwereyns, J., Koizumi, M., Kobayashi, S., & Hikosaka, O. (2001). A code for behavioral inhibition on the basis of color, but not motion, in ventrolateral prefrontal cortex of macaque monkey. *Journal of Neuroscience, 21,* 4801–4808.

Sato, M., & Hikosaka, O. (2002). Role of primate substantia nigra pars reticulata in reward-oriented saccadic eye movement. *Journal of Neuroscience, 22,* 2363–2373.

Satoh, T., Nakai, S., Sato, T., & Kimura, M. (2003). Correlated coding of motivation and outcome of decision by dopamine neurons. *Journal of Neuroscience, 23,* 9913–9923.

Schopenhauer, A. (1999). *Prize Essay on the Freedom of the Will.* (E. F. J. Payne, Transl.) Cambridge, U.K.: Cambridge University Press. (Original work first published 1839.)

Schultz, W., Apicella, P., & Ljungberg, T. (1993). Responses of monkey dopamine neurons to reward and conditioned stimuli during successive steps of learning a delayed response task. *Journal of Neuroscience, 13,* 900–913.

Schultz, W., Dayan, P., & Montague, P. R. (1997). A neural substrate of prediction and reward. *Science, 275,* 1593–1599.

Schummers, J., Yu, H., & Sur, M. (2008). Tuned responses of astrocytes and their influence on hemodynamic signals in the visual cortex. *Science, 320,* 1638–1643.

Schweighofer, N., Tanaka, S. C., & Doya, K. (2007). Serotonin and the evaluation of future rewards: Theory, experiments, and possible neural mechanisms. *Annals of the New York Academy of Sciences, 1104,* 289–300.

Sealfon, S. C., & Olanow, C. W. (2000). Dopamine receptors: From structure to behavior. *Trends in Neurosciences, 23, Supplement 10,* S34–S40.

Sebald, W. G. (2002). *Austerlitz.* (A. Bell, Transl.) London, U.K.: Penguin. (Original work published 2001; translation first published 2001.)

Seyfarth, R. M., Cheney, D. L., & Marler, P. (1980). Monkey responses to three different alarm calls: Evidence of predator classification and semantic communication. *Science, 210,* 801–803.

Seymour, B., O'Doherty, J. P., Koltzenburg, M., Wiech, K., Frackowiak, R., Friston, K., & Dolan, R. (2005). Opponent appetitive–aversive neural processes underlie predictive learning of pain relief. *Nature Neuroscience, 8,* 1234–1240.

Shadlen, M. N., Britten, K. H., Newsome, W. T., & Movshon, J. A. (1996). A computational analysis of the relationship between neuronal and behavioral responses to visual motion. *Journal of Neuroscience, 16,* 1486–1510.

Sharot, T., Davidson, M. L., Carson, M. M., & Phelps, E. A. (2008). Eye movements predict recollective experience. *PLoS ONE, 3,* e2884.

Sharot, T., Riccardi, A. M., Raio, C. M., & Phelps, E. A. (2007). Neural mechanisms mediating optimism bias. *Nature, 450,* 102–105.

Sharpee, T. O., Sugihara, H., Kurgansky, A. V., Rebrik, S. P., Stryker, M. P., & Miller, K. D. (2006). Adaptive filtering enhances information transmission in visual cortex. *Nature, 439,* 936–942.

Sherrington, C. (1955). *Man on His Nature.* Middlesex, U.K.: Penguin Books. (Original work published 1940.)

Sherrington, C. (1961). *The Integrative Action of the Nervous System.* New Haven, CT: Yale University Press. (Original work published 1906.)

Shidara, M., & Richmond, B. J. (2002). Anterior cingulate: Single neuronal signals related to degree of reward expectancy. *Science, 296,* 1709–1711.

Shimo, Y., & Hikosaka, O. (2001). Role of tonically active neurons in primate caudate in reward-oriented saccadic eye movement. *Journal of Neuroscience, 21,* 7804–7814.

Shimojo, S., Simion, C., Shimojo, E., & Scheier, C. (2003). Gaze bias both reflects and influences preference. *Nature Neuroscience, 12,* 1317–1322.

Shuler, M. G., & Bear, M. F. (2006). Reward timing in the primary visual cortex. *Science, 311,* 1606–1609.

Simon, H. A. (1956). Rational choice and the structure of the environment. *Psychological Review, 63,* 129–138.

Simons, D. J., & Rensink, R. A. (2005). Change blindness: Past, present, and future. *Trends in Cognitive Sciences, 9,* 16–20.

Singer, P. (1983). *The Expanding Circle: Ethics and Sociobiology.* Oxford, U.K.: Oxford University Press.

Singer, P. (2000). *Writings on an Ethical Life.* New York, NY: HarperCollins.

Singer, T., Seymour, B., O'Doherty, J. P., Stephan, K. E., Dolan, R. J., & Frith, C. D. (2006). Empathic neural responses are modulated by the perceived fairness of others. *Nature, 439,* 466–469.

Skinner, B. F. (1974). *About Behaviorism.* New York, NY: Alfred A. Knopf.

Small, D. M., Gregory, M. D., Mak, Y. E., Gitelman, D., Mesulam, M. M., & Parrish, T. (2003). Dissociation of neural representation of intensity and affective valuation in human gustation. *Neuron, 39,* 701–711.

Smith, A. D., & Bolam, J. P. (1990). The neural network of the basal ganglia as revealed by the study of synaptic connections of identified neurones. *Trends in Neurosciences, 13,* 259–265.

Smith, M. A., Kelly, R. C., & Lee, T. S. (2007). Dynamics of response to perceptual pop-out stimuli in macaque V1. *Journal of Neurophysiology, 98,* 3436–3449.

Smith, P. L., & Ratcliff, R. (2004). Psychology and neurobiology of simple decisions. *Trends in Neurosciences, 27,* 161–168.

Solomon, R. L., & Corbit, J. D. (1974). An opponent-process theory of motivation. I. Temporal dynamics of affect. *Psychological Review, 81,* 119–145.

Sommer, M. A., & Wurtz, R. H. (2002). A pathway in primate brain for internal monitoring of movements. *Science, 296,* 1480–1482.

Spinoza, B. (2001). *Ethics.* (W. H. White, Transl.) Hertfordshire, U.K.: Wordsworth Editions. (Original work published 1677.)

Steels, L. (2003). Intelligence with representation. *Philosophical Transactions of the Royal Society of London, A, 361,* 2381–2395.

Stein, G. (2008). *Selections.* (J. Retallack, Ed.) Berkeley, CA: University of California Press.

Sternberg, S. (1969a). The discovery of processing stages: Extensions of Donders' method. *Acta Psychologica, 30,* 276–315.

Sternberg, S. (1969b). Memory scanning: Mental processes revealed by reaction-time experiments. *American Scientist, 57,* 421–457.

Stevens, W. (1984). *Collected Poems.* London, U.K.: Faber and Faber. (Original work published 1954.)

Stigler, S. M. (1982). Thomas Bayes's Bayesian inference. *Journal of the Royal Statistical Society. Series A (General), 145,* 250–258.

Stigler, S. M. (1983). Who discovered Bayes's theorem? *The American Statistician, 37,* 290–296.

Stolz, J. A. (1996). Exogenous orienting does not reflect an encapsulated set of processes. *Journal of Experimental Psychology: Human Perception and Performance, 22,* 187–201.

Stroop, J. R. (1992). Studies of interference in serial verbal reactions. *Journal of Experimental Psychology: General, 121,* 15–23. (Original work published 1935.)

Stuphorn, V., Taylor, T. L., & Schall, J. D. (2000). Performance monitoring by the supplementary eye field. *Nature, 408,* 857–860.

Sugrue, L. P., Corrado, G. S., & Newsome, W. T. (2004). Matching behavior and the representation of value in the parietal cortex. *Science, 304,* 1782–1787.

Summerfield, C., & Koechlin, E. (2008). A neural representation of prior information during perceptual inference. *Neuron, 59,* 336–347.

Susskind, J. M., Lee, D. H., Cusi, A., Feiman, R., Grabski, W., & Anderson, A. K. (2008). Expressing fear enhances sensory acquisition. *Nature Neuroscience, 11,* 843–850.

Swets, J. A. (1961). Is there a sensory threshold? *Science, 134,* 168–177.

Swets, J. A. (1973). The relative operating characteristic in psychology: A technique for isolating effects of response bias finds wide use in the study of perception and cognition. *Science, 182,* 990–1000.

Swets, J. A. (1992). The science of choosing the right decision threshold in high-stakes diagnostics. *American Psychologist, 47,* 522–532.

Sylvester, C. M., Jack, A. I., Corbetta, M., & Shulman, G. L. (2008). Anticipatory suppression of nonattended locations in visual cortex marks target location and predicts perception. *Journal of Neuroscience, 28,* 6549–6556.

Takikawa, Y., Kawagoe, R., & Hikosaka, O. (2002). Reward-dependent spatial selectivity of anticipatory activity in monkey caudate neurons. *Journal of Neurophysiology, 87,* 508–515.

Talbot, W. H., Darian-Smith, I., Kornhuber, H. H., & Mountcastle, V. B. (1968). The sense of flutter-vibration: Comparison of the human capacity with response patterns of mechanoreceptive afferents from the monkey hand. *Journal of Neurophysiology, 31,* 301–334.

Tang, S., & Guo, A. (2001). Choice behavior of Drosophila facing contradictory visual cues. *Science, 294,* 1543–1547.

Ter Balkt, H. H. (2000). *In de waterwingebieden. Gedichten 1953–1999.* Amsterdam, The Netherlands: De Bezige Bij.

Theeuwes, J. (1994). Endogenous and exogenous control of visual selection. *Perception, 23,* 429–440.

Thompson, D. (Ed.) (1996). *The Pocket Oxford Dictionary of Current English.* Oxford, U.K.: Clarendon Press.

Thorndike, E. L. (1898). *Animal Intelligence: An Experimental Study of the Associative Processes in Animals* (Psychological Review, Monograph Supplements, No. 8). New York, NY: Macmillan.

Tipper, S. P. (1985). The negative priming effect: Inhibitory priming by ignored objects. *Quarterly Journal of Experimental Psychology, A, 37,* 571–590.

Tobler, P. N., Fiorillo, C. D., & Schultz, W. (2005). Adaptive coding of reward value by dopamine neurons. *Science, 307,* 1642–1645.

Tomita, H., Ohbayashi, M., Nakahara, K., Hasegawa, I., & Miyashita, K. (1999). Top–down signal from prefrontal cortex in executive control of memory retrieval. *Nature, 401,* 699–703.

Torralba, A., Oliva, A., Castelhano, M. S., & Henderson, J. M. (2006). Contextual guidance of eye movements and attention in real-world scenes: The role of global features in object search. *Psychological Review, 113,* 766–786.

Treisman, A., & Gelade, G. (1980). A feature-integration theory of attention. *Cognitive Psychology, 12,* 97–136.

Treisman, A., & Sato, S. (1990). Conjunction search revisited. *Journal of Experimental Psychology: Human Perception and Performance, 8,* 194–214.

Treisman, A., & Souther, J. (1985). Search asymmetry: A diagnostic for preattentive processing of separable features. *Journal of Experimental Psychology: General, 114,* 285–310.

Tremblay, L., & Schultz, W. (1999). Relative reward preference in primate orbitofrontal cortex. *Nature, 398,* 704–708.

Treue, S., & Martínez Trujillo, J. C. (1999). Feature-based attention influences motion processing gain in macaque visual cortex. *Nature, 399,* 575–579.

Tsuda, I. (2001). Toward an interpretation of dynamic neural activity in terms of chaotic dynamical systems. *Behavioral and Brain Sciences, 24,* 793–810.

Tversky, A., & Kahneman, D. (1973). Availability: A heuristic for judging frequency and probability. *Cognitive Psychology, 5,* 207–232.

Tversky, A., & Kahneman, D. (1982). Judgements of and by representativeness. In: D. Kahneman, P. Slovic, & A. Tversky (Eds.), *Judgement under Uncertainty: Heuristics and Biases* (pp. 84–98). Cambridge, U.K.: Cambridge University Press.

Ungerleider, L. G., & Mishkin, M. (1982). Two cortical visual systems. In: D. J. Ingle, M. A. Goodale, & R. J. W. Mansfield (Eds.), *Analysis of Visual Behavior* (pp. 549–586). Cambridge, MA: MIT Press.

Van der Heijden, A. H. C. (1992). *Selective Attention in Vision.* London, U.K.: Routledge.

Van der Veen, F. M., Evers, E. A., Deutz, N. E., & Schmitt, J. A. (2007). Effects of acute tryptophan depletion on mood and facial emotion perception related brain activation and performance in healthy women with and without a family history of depression. *Neuropsychopharmacology, 32,* 216–224.

Verhelst, P. (2008). *Nieuwe sterrenbeelden.* Amsterdam, The Netherlands: Prometheus.

Verhoef, B. E., Kayaert, G., Franko, E., Vangeneugden, J., & Vogels, R. (2008). Stimulus similarity–contingent neural adaptation can be time and cortical area dependent. *Journal of Neuroscience, 28,* 10631–10640.

Waelti, P., Dickinson, A., & Schultz, W. (2001). Dopamine responses comply with basic assumptions of formal learning theory. *Nature, 412,* 43–48.

Wallis, J. D., Anderson, K. C., & Miller, E. K. (2001). Single neurons in prefrontal cortex encode abstract rules. *Nature, 411,* 953–956.

Walter, W. G. (1950). An imitation of life. *Scientific American, 2,* 42–45.

Walton, M. E., Bannerman, D. M., Alterescu, K., & Rushworth, M. F. S. (2003). Functional specialization within medial frontal cortex of the anterior cingulate for evaluating effort-related decisions. *Journal of Neuroscience, 23,* 6475–6479.

Wang, Q., Cavanagh, P., & Green, M. (1994). Familiarity and pop-out in visual search. *Perception & Psychophysics, 56,* 495–500.

Washburn, D. A. (1994). Stroop-like effects for monkeys and humans: Processing speed or strength of association? *Psychological Science, 5,* 375–379.

Watanabe, K., Lauwereyns, J., & Hikosaka, O. (2003a). Effects of motivational conflicts on visually elicited saccades in monkeys. *Experimental Brain Research, 152,* 361–367.

Watanabe, K., Lauwereyns, J., & Hikosaka, O. (2003b). Neural correlates of rewarded and unrewarded eye movements in the primate caudate nucleus. *Journal of Neuroscience, 23,* 10052–10057.

Watanabe, M. (1996). Reward expectancy in primate prefrontal neurons. *Nature, 382,* 629–632.

Weinberger, N. M. (1995). Retuning the brain by fear conditioning. In: M. S. Gazzaniga (Ed.), *The Cognitive Neurosciences* (pp. 1071–1090). Cambridge, MA: MIT Press.

Weinberger, N. M. (2004). Specific long-term memory traces in primary auditory cortex. *Nature Reviews Neuroscience, 5,* 279–290.

Weinberger, N. M., Imig, T. J., & Lippe, W. R. (1972). Modification of unit discharges in the medial geniculate nucleus by click–shock pairing. *Experimental Neurology, 36,* 46–58.

Williams, C. C., Henderson, J. M., & Zacks, R. T. (2005). Incidental visual memory for targets and distractors in visual search. *Perception & Psychophysics, 67,* 816–827.

Wills, T. J., Lever, C., Cacucci, F., Burgess, N., & O'Keefe, J. (2005). Attractor dynamics in the hippocampal representation of the local environment. *Science, 308,* 873–876.

Winston, J. S., Gottfried, J. A., Kilner, J. M., & Dolan, R. J. (2005). Integrated neural representations of odor intensity and affective valence in human amygdala. *Journal of Neuroscience, 25,* 8903–8907.

Wittgenstein, L. (2001). *Tractatus Logico-Philosophicus.* (D. F. Pears & B. McGuiness, Transl.) London, U.K.: Routledge Classics. (Original work published 1921; translation first published 1961.)

Wittgenstein, L. (2003). *Philosophical Investigations.* (G. E. M. Anscombe, Transl.) Oxford, U.K.: Blackwell Publishing. (Original work published 1953.)

Wolfe, J. M. (2001). Asymmetries in visual search: An introduction. *Perception & Psychophysics, 63,* 381–389.

Womelsdorf, T., Fries, P., Mitra, P. P., & Desimone, R. (2006). Gamma-band synchronization in visual cortex predicts speed of change detection. *Nature, 439,* 733–736.

Womelsdorf, T., Schoffelen, J. M., Oostenveld, R., Singer, W., Desimone, R., Engel, A. K., & Fries, P. (2007). Modulation of neuronal interactions through neuronal synchronization. *Science, 316,* 1609–1612.

Wurtz, R. H. (1969). Visual receptive fields of striate cortex neurons in awake monkeys. *Journal of Neurophysiology, 32,* 727–742.

Zeitler, M., Fries, P., & Gielen, S. (2008). Biased competition through variations in amplitude of gamma-oscillations. *Journal of Computational Neuroscience, 25,* 89–107.

Zhang, H. H., Zhang, J., & Kornblum, S. (1999). A parallel distributed processing model of stimulus–stimulus and stimulus–response compatibility. *Cognitive Psychology, 38,* 386–432.

Zink, C. F., Tong, Y., Chen, Q., Bassett, D. S., Stein, J. L., & Meyer-Lindenberg, A. (2008). Know your place: Neural processing of social hierarchy in humans. *Neuron, 58,* 273–283.

Index

About Behaviorism (Skinner), 52
Aboutness, 155
Abstract rule, 185, 188, 192
Acetylcholine, 104
Active Boundaries (Palmer), 171
Additive scaling, 38–39, 47, 66, 72, 82, 104, 113, 143, 174, 201
Aesthetics, 226–227, 229
Alarm call, 94
Altruism, 88. *See also* Punishment, altruistic
Ambiguous figure, 161, 163
Amygdala, 59, 100, 103–104, 110, 115, 118, 132, 211, 216
Anatomy of Melancholy, The (Burton), 119
Anger, 117
Animal model, 89, 220–221. *See also* Rats versus monkeys
Answer space, 124, 174
Anterior cingulate cortex (ACC), 59, 203, 209, 211
Anticipatory processing, 38–39, 43, 47, 64–68, 72, 74–79, 82, 93, 104, 109, 113, 158, 161, 164, 167–168, 175, 201, 205–212, 215. *See also* Spike rate, before stimulus presentation
Appetitive stimulus, 103
Approach, 56, 84
Arousal, 63, 74, 76, 97, 99–100, 102
Attachment, 117
Attention, vii–ix, xi, 83, 138, 159, 162, 173, 177
  confounded with reward expectation, 81–82
  filter model of, xviii
  neural correlate of, 40
  overt versus covert, 141
  spotlight metaphor of, viii, xi, 175
  vague meaning of the word, xiv, 82, 108–109, 220, 228
Auditory discrimination, 110–115
Automatic processes, 107, 132, 137, 182
Availability, 135–137
Average, 26–27, 32–35

Aversive stimulus, 103–104
Avoidance, 7, 56, 101, 225
Aygi, Gennady, 231

Bacon, Francis, 4, 212
Bacon, Roger, 28
Badiou, Alain, 145–146, 159
Barthes, Roland, 145–146
Basal ganglia, 60, 64–65, 72, 75, 104, 211. *See also* Caudate nucleus; Dorsal striatum; Globus pallidus; Substantia nigra; Subthalamic nucleus; Ventral striatum
Bayes, Thomas, 8–9, 15
Bayesian inference, 14, 47, 142
Bayes's theorem, 8–13, 21–22, 40
Behaviorism, 52–53
Belief, 5, 85–86, 127
Beyond the Pleasure Principle (Freud), 50, 84
Bias, xiv–xv, xvii, 7, 15, 55, 85, 101, 108–110, 115, 118–120, 124–125, 133–135, 144, 157, 164–165, 167, 174–175, 197, 201, 210–211, 214, 219–222, 224–226
  for bias, 177–181
  counteractive (*see* Opponent processes)
  toward familiarity, 116, 126–130, 133, 136–137, 143, 146, 170–171
  hindsight, 125
  toward meaningfulness, 98–99, 157
  neural signature of, 38–39
  as observed in response time, 33, 35
  observer, 14, 24
  toward proximity, 153–154, 157, 162–163, 168, 170–171, 177, 180
  rational, 14
  researcher's, 211, 213
  reward-oriented, 69, 74–77, 81, 102, 187, 205–209, 211, 215
  rogue, 202
  versus sensitivity, ix, 20, 31, 37, 79, 82, 108, 139, 143, 161, 165, 168, 186, 206, 218, 228

256 Index

Bias (cont.)
  structural, 113, 132
  systems, 177–181, 183–184, 188, 195
  ways to implement, 26, 73
Bible, 4, 50
Bicuculline, 62
Biederman, Irving, 139–141, 154
Bifurcation, 162
Blanchot, Maurice, 50
Blake, William, 92
Blood oxidation level dependent (BOLD) response, 41, 87–89, 104, 118, 143, 154, 158, 160, 175, 183, 203, 215–216
Body representation, 56–57
Book of Disquiet, The (Pessoa), 96
Borges, Jorge Luis, xviii, 135
Bowlby, John, 117, 119
Broadbent, Donald, xvii–xviii, 174–175
Brain map, 59
Brainstem, 69–70
Buildup neurons, 43
Burton, Robert, 119

Carpenter, R. H. S., x, 30–31, 40, 47, 68, 134
Cartesian dualism, 29, 52
Category boundary, 203
Caudate nucleus, 59, 64–77, 80–81, 87, 102, 118, 206–210, 215
Ceiling effect, 46
Celan, Paul, 1–2, 6–7
Centromedian nucleus, 208–210
Cerebral blood flow, 41–42, 214–215
Change blindness, 153–154, 177–178
Chomsky, Noam, 3, 57, 124
Classical conditioning, 61, 103, 112
Clinton, Hillary Rodham, 129
Cogito ergo sum (Descartes), 5, 125, 219–220
Cognitive deficit, 158–159
Cognitive processes, 29, 53–54, 63, 85–86, 121, 137, 140, 177, 196, 201, 204. *See also* Information processing
Cohen, Jonathan D., 61, 190–191
Communication, 94
Competition, 177, 181, 196
  independent versus interactive, 176, 186–187
  parallel, 44
  toy model of neural, 190–195
Conceptual minimalism, xiv, 53
Conditioned reflex, 106–107, 112
Confabulation, 152–153
Confidence, 204
Conflict
  motivational, 205–206
  paradigm, 182–183, 187 (*see also* Stroop test)
  signal, 189–190, 202–204, 209

Confusability, 114
Congruency. *See* Stimulus-response compatibility
Connectionism, 131, 190
Consciousness, xvii, 63, 121, 132, 143, 197, 231
  and attention, vii
  gateway to, viii
  and language, 50, 196, 204
Consciousness Explained (Dennett), 69, 196
Conservative attitude, 115–116. *See also* Decision making, conservative
Constancy. *See* Stability
Context, 65–66, 72, 75, 83, 105, 107, 115–116, 133–134, 137, 139, 141–143, 162, 187, 214, 225
Control condition, 18, 181–182
Cooperation, 88, 101
Correct rejection, 20–23
Corticostriatal synapse, 75
Counteractive processes. *See* Opponent processes
Creativity, 124, 141, 159, 163, 171, 225–226, 229
Criterion, 21, 23–25, 33, 201. *See also* Threshold
  shift, 24–26, 39
Critical flicker frequency, 165

Damasio, Antonio R., 57, 86
Darwin, Charles, 84, 93–95
Darwinian processes, 54–55, 85, 99, 201
Darwin's Dangerous Idea (Dennett), 55, 84–85
Deadline, 101
Decision making, 19, 71, 146, 171, 197, 224, 228
  adaptive, 184
  under conflict, 187, 190, 202, 205
  conservative, 24–25
  effort-related, 209
  elaborate, 196
  error in, 36 (*see also* Probability, of error)
  liberal, 24–25
  moral, 216–218, 221
  neural model of, 8
  rational, 24
  rewriting the process of, 128
Decision line. *See* LATER model
Decision rule, 24, 27
Decision signal. *See* LATER model
Decision unit, 44
Defensive behavior, 115
DeLanda, Manuel, xii
Deleuze, Gilles, xii, 92, 122, 159, 169–170, 220
Dennett, Daniel C., 55, 69, 85, 153, 155, 195–196

Depression, 119–120
Descartes, René, 5, 57, 86, 125, 155, 219–220, 229
Desire, 50, 64–65, 200, 202
Determinist model, 160
De Waal, Frans B. M., 101, 221
Dickinson, Emily, 173
Difference, 122, 146–147, 163, 170
Difference and Repetition (Deleuze), 169–170
Dimensional overlap theory, 195
Direct versus indirect pathway, 71
Disgust, 93
Disinhibition, 62, 72–73
Distraction, viii, 9
Donders, Franciscus Cornelis, 28–29
Dopamine, 60, 63, 72, 78, 82, 84, 211. *See also* Prediction, error
  neural activity, 61–63, 65, 69, 75, 102, 223
  neurotransmitter, 61, 75, 101, 209
  projection, 71, 74, 79, 81, 83, 86
Dorsal striatum, 64, 68, 74–75, 81, 87. *See also* Caudate nucleus
Dorsolateral prefrontal cortex (DLPFC), 59–60, 78, 102–103, 185, 188–195, 205, 216
Double standards, 128
Dragomoshchenko, Arkadii, 143
Dream analysis, 105
Duhem, Pierre, 160–161
Dynamics, xii, 79, 141, 155, 158, 161, 225
  attractor, 159–160, 162–163, 170–171
  behavioral, 55
  of bias and sensitivity, 7
  of synaptic weights, 74, 105, 110, 113–114, 168, 209
  temporal, 45, 78, 81, 162
  of thought, vii, xi

Emotion, 57, 63, 86, 93–94, 216–219
Empathy, 215–216, 220–221
Episodic coding, 105–106, 138, 153
Epithalamus, 102, 211
Equality, 128–129, 169
Equity, 101, 217
Ergodic hypothesis, 31
Escape. *See* Avoidance
Eternal return, 92, 107, 155, 169
Ethics, 41–42, 129, 213, 220–222, 227
Ethics (Spinoza), 226
Event, 145–147, 162
Event-related potential, 146
Evolutionary model, 55. *See also* Darwinian processes
Expanding Circle, The (Singer), 129
Expectation, 85, 145, 147, 170. *See also* Reward, expectation
  guiding interpretation, 3, 7
Experimental condition, 18
Exploration, 115, 127, 143, 175

Exponential distribution, 17–18
Expression of the Emotions in Man and Animals (Darwin), 94–95
Eye movement, 15, 50, 100, 164, 205–208
  anticipatory, 143
  control, 23, 31, 69–74, 138, 209–210
  memory-guided, 64–65, 103
  visually guided, 66–68

Fairness, 14, 125, 215–218, 221
False alarm, 20–25, 51–52, 77, 98, 109–110, 120, 126, 128, 137, 140
False memory, 153
Falsifiability, 212
Familiarity, 122, 125–131, 142, 144–145. *See also* Bias, toward familiarity; Perceptual grouping, by familiarity
Faverey, Hans, 142
Fear, 91–99, 103, 110, 200, 202
  conditioning, 104–105, 110–116
  of snakes, 105–109, 115
  and sorrow without a cause, 119–120
  of the unknown, 116–117
Feature-integration theory, 30, 108, 145, 162–163
Feelings, 57, 63
Field potential, 160
Firing rate. *See* Spike rate
Flat taxonomy, 134, 136
Flexibility, 204, 220
Functional magnetic resonance imaging (fMRI), 42, 89, 100, 120, 132, 137, 168, 215–217, 220
  of conflict, 190
  of fear-related activity, 104–105, 112
  of object-selective activity, 157
  of orientation discrimination, 43
  of reward-related activity, 76, 81, 87
Forced choice, 20–21, 185, 203
Foucault, Michel, 135, 163
Framing effect, 118
Freud, Sigmund, 50, 52, 105–106
Frontal eye field (FEF), 23, 59, 69–74, 78–80, 83, 185, 188–189, 191–194, 203, 210
Fruit flies
  facing contradictory cues, 183
  on speed, 89, 102

Gain, 22, 44, 117–118, 161. *See also* Multiplicative scaling
Gamma-aminobutyric acid (GABA), 62, 69, 71, 75, 104, 206
Gamma band, 161–162
Game theory, 213
Gaussian distribution, 17
Gene-culture coevolution, 214
Genetic endowment, 53–54, 115

Geniculate nucleus
  lateral (LGN), 59, 138
  medial, 113
Gestalt, 167, 171
  psychology, 147, 155, 158–159, 162–163, 171
Gibson, J. J., 122, 146, 174
Gigerenzer, Gerd, 9, 12, 98
Glass ceiling, 129
Glimcher, Paul W., 40, 88
Globus pallidus
  external segment, 71
  internal segment, 211
Glutamate, 69, 75, 104
Goldberg, Michael E., 40, 138, 164, 177, 231
Grammar of thinking, 85
Gravity of harmony, 161, 225, 228
Green, David M., 19–21, 134
Guattari, Félix, 159, 220

Happily (Hejinian), xiii–xiv
Happiness, 63, 83–86, 95–96, 100
Harmonium (Stevens), 121
Hauser, Marc D., 3, 57
Hebb, Donald Olding, 113, 174
Hedonics, 74–75, 84, 100
Heidegger, Martin, 219–220, 225
Hejinian, Lyn, x–xi, xiii–xiv, xviii
Herrnstein, Richard J., 88
Heuristic, 126, 133, 135–137, 179
Hierarchical structure, 3, 83, 138
Hikosaka, Okihide, 40, 64–66, 68–69, 75–77, 102, 143, 178, 206–207, 231
Hirohito, Emperor, 152
Hit, 20–23, 140
Hitler, Adolf, 123–125
Hofstadter, Douglas R., 57
Hölderlin, Friedrich, 7
Holocaust, 1
Homeostasis, 218
Hope, 85, 97–99, 104
Howe, Susan, 199, 222
Hume, David, ix, 77–78, 97, 205

Ideal observer, 20
Ideal of learning, 119
Identity, 169. *See also* Subjectivity
Imagination, 170
Implicit Association Test, 131–133
Inattentional blindness, 178–179
Incentive value, 65–66
Inclusive fitness, 55–57, 85, 96, 201
Individuality. *See* Subjectivity
Inequity. *See* Equity
Infinite regress, 5, 84, 158
Information processing, 85, 133, 137, 142, 146, 171, 178, 186–187, 205. *See also* Cognitive processes
  adaptive control of, 184, 190
  bottom-up versus top-down, 138–139, 142, 144, 162 (*see also* Top-down control)
  facilitation versus inhibition of, viii, 82, 145, 175, 182, 185, 206, 209
  irrelevant, 188–189, 193–195
  parallel, 29–30, 108, 144, 162 (*see also* Parallel competition)
  selective, 154, 175, 228
  serial, 29–30, 145
  ventral versus dorsal stream of visual, 187, 191–195
Informativeness, 141–142
Inhibitory neuron, 62, 69–70, 80, 102, 209–210
Injustice. *See* Justice
Innovation, 116
Instrumental learning. *See* Operant conditioning
Interference, 182–187, 190–191, 195
Internal state, 56–57
Intracellular mechanism, 74, 168
Intracellular recording, 74
Intracranial self-stimulation (ICCS), 74–75

James, William, vii–xii, xiv, xviii, 55, 109, 163, 174, 228
Joy. *See* Pleasure
Justice, 53, 124, 129, 136, 215, 217

Kahneman, Daniel, 88, 117, 133, 135–137, 179
Keats, John, 49, 120, 226
Kenkō, xiii, 14, 51
King, Martin Luther, 129
Knowledge acquisition, 6

Lacan, Jacques, 50–53, 57, 224
Language, x–xi, xiv, 2, 51, 57, 98, 124, 129–130, 147, 157, 183, 229–230. *See also* Consciousness, and language
  and desire, 50, 52
  ontogeny of, 94
  and shared experience, 122
Language of Inquiry, The (Hejinian), x
Lashley, Karl S., 58, 83
Lateral geniculate nucleus (LGN). *See* Geniculate nucleus, lateral
Lateral habenula, 102, 211
Lateral intraparietal cortex (LIP), 43, 59, 69, 164
LATER model, 30–37, 39–40, 43–44, 47, 68–69, 72–73, 78–80, 101, 110, 114, 128, 134, 141, 161, 165, 174, 186, 201, 224
  and change in rate of processing, 34–35, 46, 80
  and change in starting point, 33, 35, 73, 93, 157, 167
  with rats, 68

Law of Effect, 56
Law of Eponymy, 9, 40
Laziness, 101, 151
Leadership, 127
LeDoux, Joseph, 110
Less is more, 171, 173–174, 175, 177–181
Life regulation, 56–57
Likelihood. *See* Probability
Liking versus wanting, 64
Location parameter, 17, 27
Logical inference, 87
Logic of Scientific Discovery, The (Popper), 212
Long-term potentiation and depression, 74
Looking for Spinoza (Damasio), 57
Lorenz, Konrad, 106
Loss aversion, 117–118, 218
Luce, R. Duncan, 19, 30, 35

MacIntyre, Alistair, 83
Malaria, 229–230
Man on his Nature (Sherrington), 229–230
Marr, David, 138–139
Marriage of Heaven and Hell (Blake), 92
Mass Action in Cerebral Function (Lashley), 58
Matching Law, 88
Maunsell, John H. R., 81–82
Mean. *See* Average
Meaningfulness, 98, 200. *See also* Bias, toward meaningfulness
Medial prefrontal cortex, 216
Meme, 130
Memory search, 29–30
Mesocortical pathway, 81
Metacognition, 204
Middle fusiform gyrus, 143
Middle temporal area (MT), 59, 81, 191–194
Milton, John, 91–92, 119
Minimalist theorist, 5, 98–99, 154
Miss, 20–25, 98
Moral judgment, 106, 217, 220–221, 227. *See also* Decision making, moral
More, Thomas, 199–200
Mosquito, 229–230
Motivation, 64, 219. *See also* Conflict, motivational
Mountcastle, Vernon B., 37
Multiplicative scaling, 44–47, 78–79, 81–82, 104, 142, 161–162, 186, 201
Mushroom bodies, 183

Naming, 146–147
Narrative technique, 3
Natural selection, 84, 96. *See also* Darwinian processes
Negative priming, 165

Neural adaptation, 164–165
Neural circuit, 69–74, 79–80, 83, 96, 104–105, 107, 113, 115, 130–131, 138, 153, 170, 187, 190–195, 204, 209–211
Neural modeling, 62
Neural overlap theory, 195
Neuroeconomics, 40, 87–88
New Organon, The (Bacon), 4, 212
Nietzsche, Friedrich, 92, 107, 169
Nigrostriatal pathway, 81
Noise. *See* Signal-to-noise ratio
Nonlinearity, 79, 170–171, 186
Normal distribution, 17–18, 21
Nose poke, 68–69
Novelty, 145
Nucleus accumbens, 216. *See also* Ventral striatum
Nukata, Princess, 49–52

Obama, Barack, 128, 147
Objectivity, 6, 219, 222
Object recognition, 122, 138–139, 146–154–155, 157, 161
Objects or trains of thought, vii, xi, 2, 159, 174, 180
Obsessive-compulsive ideation, 7
Occam's razor, 180, 221
Ode on Melancholy (Keats), 120
Odor discrimination, 105, 203
Oedipus, 3
Open Society and Its Enemies, The (Popper), 212–213
Operant conditioning, 56, 61, 103
Opponent processes, 102, 197, 202, 206–212, 221
Optimism, 211–212
Orbitofrontal cortex (OFC), 59, 87, 100, 203
Order of Things, The (Foucault), 135
Organization of Behavior, The (Hebb), 113
Orpheus and Eurydice, 50
Oscillation, 160, 162
Oval Window, The (Prynne), 90

Pain, 95–96, 103, 205, 216, 218
Palmer, Michael, 46, 115–116, 158, 171
Pandemonium, 196
Paradise Lost (Milton), 91–92
Parafascicular nucleus, 208–210
Parallel search. *See* Information processing, parallel
Parietal cortex, 83, 158–159. *See also* Lateral intraparietal cortex; Middle temporal area (MT)
Pattern recognition, 157–158, 162. *See also* Object recognition
Pavlov, Ivan, 106
Pavlovian conditioning. *See* Classical conditioning

Perceptual grouping, 156, 159, 165, 169–171, 224
  by familiarity, 147, 157
  by proximity, 163
  by similarity, 144–145, 147, 155, 157
Perceptual load, 175
Perceptual organization. *See* Perceptual grouping
Performance monitoring, 190, 192, 195, 203, 205, 210. *See also* Conflict signal
Peri-aqueductal gray (PAG), 59, 104
Personal involvement, 216–217
Perspective, 6, 150, 154, 219
Pessoa, Fernando, 96
Philosophical Investigations (Wittgenstein), 85, 179
Physiological measures, 63, 115
Pinker, Steven, 130–131, 196
Placebo effect, 18, 99, 212
Pleasure, 62–63, 84–85, 205, 215, 218, 226–227. *See also* Satisfaction
Plous, Scott, 133–135
Poetry, xi, 15, 159, 171, 173, 199, 229
  algorithms borrowed from, 124
  definition of, 2
  difficult, 1–2, 7
  etymology of, 157, 170
  experimental, xi, 149, 184
  and inventing names, 146
Popper, Karl, 3, 51, 211–213, 224
Positron emission tomography (PET), 42, 214–215
Posner, Michael I., viii, 82, 175
Posterior cingulate cortex, 216
Postsynaptic potential, 75
Prägnanz, 155, 158, 162–163
Prediction, 7, 4, 28, 104, 126, 144–145, 203–205, 212, 225, 231. *See also* Reward, prediction
  error, 60–63, 74, 88, 102, 211, 223
  of a predictor, 76
  and remapping, 164
Preference, 153
Prefrontal cortex, 83, 86–87, 104, 132, 167–168, 188. *See also* Dorsolateral prefrontal cortex (DLPFC); Frontal eye field (FEF); Medial prefrontal cortex; Orbitofrontal cortex (OFC); Ventrolateral prefrontal cortex (VLPFC)
Prejudice, 14, 128, 132–133, 146. *See also* Racism; Sexism
Premotor cortex, 100
Preparatory activity, 168
Primary auditory cortex, 110–114
Primary visual cortex (V1), 59, 62
Principles of Psychology, The (James), vii–xi
Prior. *See* Probability, prior

Prisoner's dilemma, 213
Probability, 127, 136
  conditional, 10, 12
  distribution, 15–17, 22, 25–27
  of error, 19, 23–26, 184, 194–195, 203–205
  joint, 12
  marginal, 10
  posterior, 9–11, 22
  prior, 8–10, 13, 22, 26, 39–40, 42
  of reward (*see* Reward, probability)
  theory, 7–8
Procrastination, 31
Prospective coding, 68, 76
Prospect theory, 117–118
Proximity, 136–137, 139, 147, 150–155, 162–163, 165. *See also* Bias, toward proximity
Prynne, J. H., 89–90
Psychology of Judgment and Decision Making, The (Plous), 133
Publication culture, xii–xiii, 211, 213
Punishment, 56, 100, 214. *See also* Reinforcement, negative
  altruistic, 214–216
Punning, 181

Quantitative approach, 28
Quine, W. V., 179–180

Racism, 14, 123, 132–133
Ratcliff, Roger, 36, 47, 134
Rational choice, 88
Rational ideal, 8, 213
Rationality, 86, 216–217, 227
Rats versus monkeys, 42, 221–222
Reciprobit plot, 32
Reciprocity, 88
Reckoning with Risk (Gigerenzer), 9
Recognition-by-components theory, 139, 154
Recursion, 3, 11–12, 56, 77, 99, 173, 180, 196, 224
Redgrave, Peter, 61–62
Reinforcement, 56, 60, 75, 85, 103, 107, 192
Repetition, 107, 145–147, 170
  priming, 165–169
Representativeness, 135–137
Residual activity, 166–167
Response time, 161, 184, 186, 194–195, 206–208
  correlated with spike rate, 47, 67–68
  measuring the, 16, 66, 141
  inverse of, 32, 35
  shape of the distribution of, 18, 30–35
  as a tool to study cognition, 28–29
Restlessness, 96–97, 99, 101, 104, 120
Retinal ganglion neurons, 137–138
Retrospective coding, 78
Revenge, 215–216

Reward, 56, 66–69, 83, 101–103, 205–209, 211, 218, 225. *See also* Dopamine; Prediction, error; Reinforcement, positive
  availability, 65
  behavior oriented to, 58
  expectation, 60
  factor (*see* Reward, information)
  information, 65, 71, 74, 78–79, 81, 87, 100
  magnitude, 82
  possibility of, 64
  prediction, 61, 63, 188
  probability, 61, 82, 88
Rhinal cortex, 101
Russell, Bertrand, 201

Sappho, 149
Satisfaction, 56, 62–63, 215
Scale parameter, 17, 27
Schopenhauer, Arthur, 205
Schultz, Wolfram, 60–62, 87, 103
Selection, 174–176
Self-fulfilling prophecy, 99, 212
Self-organization, 158–159, 162–163, 170–171
Semantic consistency, 139–142
Semantic network, 131–133
Semantic priming, 131, 133, 138, 144–145
Sensitivity, 26–28, 85, 93, 104–105, 108, 113, 139–140, 145, 169, 175, 186, 189. *See also* Bias, versus sensitivity
  heightened, 82, 142
  neural signature of, 44–45, 79
  as observed in response time, 34–35
  reward-dependent, 79, 81–82
  ways to increase, 27–28, 80
Sentence art, x–xi
Separation, 117
Serendipity, 145
Serial search. *See* Information processing, serial
Serotonin, 104, 120, 209
Set size, 30, 43, 144, 162
Sexism, 14, 129, 132–133
Sherrington, Charles, 229–230
Signal detection theory, 37, 39, 43–44, 47, 51, 82, 98, 128, 139–140, 161, 174, 201, 224
  applying, 66, 72, 134
  the concepts of, 20–27
Signal Detection Theory and Psychophysics (Green & Swets), 19
Signal distribution. *See* Signal-to-noise ratio
Signal-to-noise ratio, 20–27, 38–39, 44–46, 72, 79, 82, 93, 104–105, 108, 114, 142, 162, 167–168, 175, 186, 189, 203–204, 218
Signifier and signified, 98–99
Similarity, 82, 195. *See also* Perceptual grouping, by similarity

Simon, Herbert, 98, 179
Simplicity, 179–180
Singer, Peter, 129–130, 221
Single-unit recording, 19, 40–41, 43, 60, 88, 103, 160, 187, 221. *See also* Rats versus monkeys; Spike rate
Skinner, Burrhus Frederic, 52–56, 62, 64, 85, 103, 107
Social reputation, 87, 214
Social status, 87
Spatial resolution, 41–42
Spike rate, 19, 23–24, 36–37, 104, 113, 142, 160, 167, 185, 187–189, 203
  before stimulus presentation, 65, 67–68, 74
  after stimulus presentation, 206–208
Spinoza, 226
Spreading activation, 131, 133
Stability, 154, 164–165
Standard deviation. *See* Variability
Starting point. *See* LATER model
Statistics, 7–9, 18, 23, 31–32, 136
  of the environment, 64, 127, 170
  inferential, 16
  as a metaphor, 8
  nonparametric, 13
Stein, Gertrude, xviii, 2, 147
Sternberg, S., 29–30
Stevens, Wallace, 121–122
Stimulus-response compatibility, 183–195, 203
Stroop, J. Ridley, 181–182
Stroop test, 181–184, 203, 205
  for monkeys, 184–195
Structure of knowledge, 135
Studium and punctum, 145–146
Subjectivity, 5–7, 13, 50, 106, 200, 218–221
Substantia nigra
  pars compacta (SNc), 59–60, 69–75, 80, 102, 210. *See also* Dopamine
  pars reticulata (SNr), 59, 69–73, 75, 78–80, 209–210
Subthalamic nucleus, 71
Subtraction method, 29
Subtractive scaling, 174–175
Superadditivity, 186–187
Superior colliculus (SC), 23, 43, 59, 62, 65, 69–75, 78–80, 83, 168, 210
Supplementary eye field (SEF), 59, 74, 143, 185, 188–195, 203, 209–210
Swets, John A., 19–21, 134
Sword of Damocles, 7
Symbolic representation, 94, 106
Symmetry, 155, 158
Synaptic plasticity. *See* Dynamics, of synaptic weights
Synchronization, 160–161
Synergistic processing, 45–47, 78–79, 81–82, 100, 168

Talking cure, 51, 106, 224
Tautology, 11–12, 51
Temporal cortex. *See also* Visual area four (V4)
  inferior, 142
  medial, 153
Temporal resolution, 41–42
Thalamus, 65, 75, 102, 110. *See also* Centromedian nucleus; Geniculate nucleus; Parafascicular nucleus
Thorndike, Edward L., 55–56, 62
Threshold, 23–24, 31–35, 40, 43, 69, 71–72, 109–110, 114, 167, 174, 201
Tonotopic organization, 111–114
Top-down control, 82–83, 137, 147, 163, 204–205
Tractatus Logico-Philosophicus (Wittgenstein), 1–2, 179
Treisman, Anne, 30, 108, 162
Triangulation, 36, 47, 69
Trust game, 87–88, 214–215
Truth
  versus goodness and beauty, 226–229
  procedure, 6–7
  timeless, 201
Tuning curve, 37–39, 44–46, 111–112, 161–162, 164, 168–169
Tversky, Amos, 88, 117, 133, 135–137, 179
Two-choice task. *See* Forced choice

Uncertainty, 26, 43, 88, 97, 125, 205
Utilitarianism, 216, 226
Utility, 118
Utopia, 133, 199–201, 205, 213, 222

Valence, 99–100, 104
Variability, 15–16, 26–27, 32–35, 109
  interindividual, 106
  random, 31
  trial-by-trial, 37, 41, 47, 55
Variance. *See* Variability
Ventral striatum, 61
Ventral tegmental area, 60, 81. *See also* Dopamine
Ventrolateral prefrontal cortex (VLPFC), 59, 185, 188–195
Viewpoint. *See* Perspective
Violence, 124, 127
Vision (Marr), 138
Visual area four (V4), 59, 81, 161, 164–166, 191–194
Visual discrimination, 66, 138, 158, 186
  of color, 165–166, 183–185, 187–195
  of motion direction, 43, 46, 184–185, 187–189, 191–195
Visual search, 15, 18, 30, 143–144, 162, 177
Voluntary control. *See* Top-down control

Weibull distribution, 17–18
Weinberger, Norman M., 110
Wernicke's area, 2
What Is Philosophy? (Deleuze & Guattari), 220
Wishful seeing, 66, 77, 79, 224, 226
Wittgenstein, Ludwig, 1–2, 85–86, 179
Wurtz, Robert H., 40–41, 62, 138, 231